Milk Quotas

MILK QUOTAS

European Community and United Kingdom Law

MICHAEL CARDWELL
Lecturer, Faculty of Law
University of Leeds

CLARENDON PRESS · OXFORD
1996

Oxford University Press. Walton Street, Oxford OX2 6DP

Oxford New York
Athens Auckland Bangkok Bombay
Calcutta Cape Town Dar es Salaam Delhi
Florence Hong Kong Istanbul Karachi
Kuala Lumpur Madras Madrid Melbourne
Mexico City Nairobi Paris Singapore
Taipei Tokyo Toronto
and associated companies in
Berlin Ibadan

Oxford is a trade mark of Oxford University Press

Published in the United States
by Oxford University Press Inc., New York

British Library Cataloguing in Publication Data
Data available

Library of Congress Cataloging in Publication Data
Data available

ISBN 0-19-825940-9

1 3 5 7 9 10 8 6 4 2

Typeset by Best-set Typesetter Ltd., Hong Kong
Printed in Great Britain
on acid-free paper by
Biddles Ltd., Guildford and King's Lynn

To my mother and father

Acknowledgements

A tremendous amount of assistance and encouragement has been received throughout the preparation of this book. Not least, with the subject matter covering both European Community and detailed United Kingdom agricultural law, it has been a great benefit to receive the comments and advice of so wide a range of experts across those fields.

At Leeds University Professor John Bell read the vast majority of the book in draft; and many improvements were instigated as a result of his suggestions. Adrian Wood provided much input on the European Community aspects; Sarah Lane (together with Tim Illston at Burges Salmon) made contributions of real substance to the chapter on taxation; and Professor Horton Rogers gave guidance on aspects of contract. At the same time a debt of gratitude is owed to all those who shouldered the extra burden during my period of study leave and to the Library Staff.

Away from Leeds University assistance was received from James O'Reilly S.C. and Oliver Ryan-Purcell in Ireland, Edwin Peel of Keble College, Oxford, Professor Chris Rodgers of the Centre for Law in Rural Areas, Aberystwyth, and Angela Sydenham of the Country Landowners Association. Further, it is difficult to overestimate the benefit of advice received from former colleagues at Burges Salmon. If the book extends beyond an academic treatment of the subject, this is in large part a tribute to them. Among many others may be specifically mentioned Andrew Densham, Della Evans, Tim Illston, Michael Johnstone, William Neville, Neil Porter, Pat Walker and Vicky Brooks of the Library Staff, and last, but by no means least, Peter Williams. Indeed, it was as an articled clerk working for Peter Williams on the case of *Puncknowle Farms Ltd.* v. *Kane* that I first encountered milk quotas; and his longstanding encouragement and numerous comments upon draft chapters have proved central to the completion of the book.

Grateful thanks are also due to the staff at Oxford University Press. The efficiency and calm with which they handled matters of production was very much appreciated, as was their willingness to accommodate recent changes in the law. Finally, the assistance of Clare Horsley with the typing led to greater speed and accuracy in the drafting of the text.

Michael Cardwell
York
January 1996

Table of Contents

Tables of Cases

TABLE OF CASES BEFORE THE EUROPEAN COURT OF JUSTICE AND THE COURT OF FIRST INSTANCE

(Numerical)

TABLE OF CASES BEFORE NATIONAL COURTS AND AUTHORITIES

European Community Legislation

DECISIONS

DIRECTIVES

International Conventions and Agreements

Table of National Legislation

Introduction

The milk quota system has now been in place for over a decade and, under the current legislation, is to endure until the year 2000.[1] Over time it has become established as a cornerstone of the dairy sector of the Common Agricultural Policy; and, indeed, the absence of a quota effectively precludes farmers from operating in that sector. None the less, milk quotas remain an elusive and esoteric subject, such characteristics being in part attributable to the highly detailed and, on occasion, highly technical regulatory framework. Moreover, there is frequently an uneasy match between Community law and national law, this difficulty being perhaps most pronounced in the application of Community provisions to landlord and tenant relationships.

Three reasons for closer consideration of the milk quota system may be highlighted. First, as indicated, the system is a major plank in the Common Agricultural Policy which continues to enjoy special treatment under the EC Treaty and to dominate the Community budget. For example, in 1995 the total Community budget amounted to 76,526.1 million ECU, of which 39,946.9 million ECU were for agriculture.[2] Further, while it would be impossible to argue that the majority of the regulations which govern the Common Agricultural Policy are of broad application, it is hard to ignore their sheer number. An illustration may be provided by the original Commission Regulation laying down the detailed rules for milk quotas.[3] Dated 16 May 1984, it was already being amended for the ninth time by 24 April 1985.[4] Besides, within that volume of legislation, several provisions have inevitably given rise to far-reaching consequences,

---

[1] The expressions 'milk quotas' and 'reference quantities' have been used interchangeably. Community legislation as a rule continues to employ the latter, but the former enjoys widespread currency — and, indeed, the term 'quota' has been adopted in UK legislation. For a discussion of the confusion which may arise, see Case 203/86, *Spain* v. *Council*, [1988] ECR 4563, at 4567–8. On a strict analysis it may be argued that there is no 'quota' as such, since the legislation does not impose an absolute prohibition on production but rather renders excess production economically unviable — see, e.g., Lawrence, G., 'Milk Superlevy: the Community System', in *Milk Quotas: Law and Practice: Papers from the ICEL Conference — June 1989* (Trinity College, Dublin, 1989), 1–10, at 1; and Gehrke, H., *The Implementation of the EC Milk Quota Regulations in British, French and German Law* (European University Institute, Florence, 1993), at 31. In this book the expression 'reference quantities' will be confined to its more technical sense of allocations to individual producers or purchasers; and, in line with common usage, the expression 'milk quotas' will be employed elsewhere — as, e.g., the 'milk quota system'.

[2] European commission, *The Agricultural Situation in the Community: 1994 Report*, at T/86.

[3] Comm. Reg. (EEC) 1371/84, [1984] OJ L132/11.

[4] Comm. Reg. (EEC) 1043/85, [1985] OJ L112/18.

for example with regard to the non-contractual liability of the Community institutions.

Secondly, it has become accepted that entitlement to a reference quantity is a *sine qua non* of profitable operation in the dairy sector. That farmers perceive the risks and potential sanctions of milk production without such entitlement to be too great may be judged from the high prices that they have been willing to pay for quota. Thus, towards the close of the 1994/5 milk year United Kingdom milk producers were reported to be paying in the region of 70 pence per litre for its permanent acquisition. Further indication is provided by the frequency of transactions. By way of illustration, some 2,500 applications for permanent transfer and 6,270 for temporary transfer were submitted in the period 1 April to 15 July 1995.[5] So radical a transformation of the dairy sector could not but have widespread economic and social consequences for producers, as foreseen by the Community institutions; and pervading the Community legislation may be found measures to address such concerns.

Thirdly, agriculture has played a large part in the development of Community law, with milk quotas making a notable contribution. Over and above the volume of regulations a significant number of decisions have emanated from both the European Court and the national courts. Indeed, over the period 1988–93 agriculture proved the source of more preliminary rulings and more direct actions than any other subject area governed by the EC Treaty;[6] and, to take an extreme example, agriculture formed the subject matter of 157 out of the 357 cases brought in 1990.[7] Besides, in 1993 such was the weight of litigation instigated by one category of milk producer that it became necessary to transfer claims to the Court of First Instance.[8] At the same time the growing body of decisions has provided scope for the development of the general principles of Community law; and, with milk quotas imposed across diverse forms of land tenure throughout the Member States, there has been ample opportunity to adjudicate upon the interaction between Community and national provisions. As a result, it may now be argued that milk quotas provide a valuable bank of material of general relevance to common organizations of markets. In support of this argument may be advanced the case of *Germany* v. *Council*, where the European Court applied its earlier decisions

---

[5] *IB Press Notice* 14/95.

[6] See, e.g., Brown, L. N. and Kennedy, T., *The Court of Justice of the European Communities* (4th edn., Sweet and Maxwell, London, 1994), at App.V, Tables 7 and 9.

[7] *Synopsis of the work of the Court of Justice and the Court of First Instance of the European Communities in 1990 and record of formal sittings in 1990* (Official Publications Office, Luxembourg, 1991), at Table 3.

[8] Decision of the Member States of 8 June 1993, [1993] OJ L144/21; and see also Brown, L. N. and Kennedy, T., *The Court of Justice of the European Communities*, at 75–6 and 167–8.

on milk quotas extensively to the common organization of the market in bananas, constituted in 1993.[9]

Against this background it is central to the purpose of this book to address both the Community and the national elements of the milk quota system. In this context emphasis is placed on the system as implemented in the United Kingdom, with its particular forms of land tenure and its particular approach to milk marketing — both of which have undergone large-scale and recent reform.[10]

In the first part of the book the aspects considered are mostly of relevance to the Community as a whole; emphasis is placed upon the allocation of milk quotas and the determination of liability on excess production. Chapter 1 is concerned with the reasons for the introduction of milk quotas, the initial legislation, and challenges to that legislation; Chapter 2 with subsequent allocations and, in particular, allocations to producers who had been excluded from reference quantities through participation in Community schemes; and Chapter 3 with the growing need for reform and the implementation of reform as from 1992. In the second part of the book, Chapters 4 and 5, there is discussion of quota transfers and landlord and tenant issues. Both aspects are to a significant extent dependent upon the agricultural law applicable in each Member State and, as indicated, the focus is upon implementation in the United Kingdom. The third part of the book, Chapter 6, addresses matters of taxation, a question of critical importance being whether or not milk quota constitutes part of the land.

---

[9] Case 280/93, [1994] ECR I–4973.

[10] For excellent comparative treatments, see, e.g., Burrell, A. M. (ed.), *Milk Quotas in the European Community* (CAB International, Wallingford, Oxfordshire, 1989); and Gehrke, H., *The Implementation of the EC Milk Quota Regulations in British, French and German Law*.

1

The Introduction of Milk Quotas

1. THE SURPLUS IN MILK AND MILK PRODUCTS

Milk quotas were introduced on 2 April 1984 to curb the endemic structural surplus in the dairy sector of the Common Agricultural Policy.[1] In the period from 1973 to 1981 milk deliveries had increased on an annual basis by approximately 2.5 per cent, the rate of increase accelerating to about 3.5 per cent in 1983.[2] However, consumption within the Community had grown at a lesser rate and, immediately prior to the introduction of milk quotas, had tended to stagnate or even decline. Accordingly, by 1983 milk deliveries were close to 104 million tonnes, while consumption stood at only 82.4 million tonnes.[3] In addition, while the Community had achieved self-sufficiency in dairy products as a whole by 1974,[4] in certain key areas the degree of over-supply had become critical: for example, taking an average over the three years to 1981, production of whole-milk powder amounted to 337 per cent of its requirements.[5]

Associated with this surplus were massive disposal costs. The extent of such costs may be judged by the Commission's estimate that in 1984 expenditure on intervention buying and export subsidies would amount to

---

[1] For Community literature on the introduction of milk quotas, see, e.g., COM(83)500; and *Bulletin of the European Communities*, No.3, 1984, at 10–14. See also *First Report from the Agriculture Committee: the Implementation of Dairy Quotas* (Session 1984–5, H.C.14). More generally, see, e.g., Avery, G., 'The Common Agricultural Policy: a Turning Point?', (1984) 21 C.M.L.Rev. 481–504; Snyder, F. G., *Law of the Common Agricultural Policy* (Sweet and Maxwell, London, 1985), at 124–9 and 142–51; Harvey, D. R., *Milk Quotas: Freedom or Serfdom?* (Centre for Agricultural Strategy, 1985), at 23–42; Wood, D., Priday, C., Moss, J. R., and Carter, D., *Milk Quotas: Law and Practice* (Farmgate Communications, 1986), at 1–7; Petit, M., de Benedictis, M., Britton, D., de Groot, M., Henrichsmeyer, W., and Lechi, F., *Agricultural Policy Formation in the European Community: the Birth of Milk Quotas and CAP Reform* (Elsevier, Amsterdam, 1987), *passim*; Usher, J. A., *Legal Aspects of Agriculture in the European Community* (Clarendon Press, Oxford, 1988), at 72–82; Lawrence, G., 'Milk Superlevy: the Community System', in *Milk Quotas: Law and Practice: Papers from the ICEL Conference — June 1989*, at 1–10; Apsion, G., *Milk Quotas* (Farm Tax and Finance Publications, 1992), *passim*; and Gehrke, H., *The Implementation of the EC Milk Quota Regulations in British, French and German Law*, at 29–40.

[2] Court of Auditors, *Special Report No.2/87 on the Quota/Additional Levy System in the Milk Sector*, [1987] OJ C266/1, at 1.1.

[3] Ibid., at Table 3.   [4] Ibid., at 1.1.

[5] European Commission, *The Agricultural Situation in the Community: 1983 Report*, at 110. Frequent reference will be made to this report, it being roughly contemporaneous with the decision to introduce milk quotas. See also Snyder, F. G., *Law of the Common Agricultural Policy*, at 124–9.

£1,565.5 million and £1,204.4 million respectively. Moreover, the total of £2,805.9 million to be spent in the dairy sector would comprise some 30.3 per cent of Common Agricultural Policy expenditure for that year.[6]

As a result, even within a Common Agricultural Policy characterized by over-production, the Commission could identify the surplus in milk and milk products as that presenting the greatest challenge.[7] The need for concerted and radical action was exacerbated by the fact that, while there had been a common organization of the market in dairy products since 1968, specific measures enacted within the framework of that legislation had failed to balance supply and demand.

2. EARLIER MEASURES TO COMBAT THE SURPLUS

The common organization of the market in dairy products was implemented pursuant to Article 43 of the EC Treaty by Council Regulation (EEC) 804/68.[8] Since then the market has been based upon a single target price for milk fixed during the currency of each milk year, the milk year generally running from 1 April to 31 March.[9] Within the Community, market support has been achieved by setting intervention prices for butter, skimmed-milk powder, and Grana Padano and Parmigiano Reggiano cheeses. The first two of those milk products have proved particularly well-suited to storage.[10] External protection, on the other hand, is achieved by setting threshold prices for certain representative pilot products, the prices of imported products being raised to the level of these threshold prices by means of a variable levy.[11] In addition, export refunds may be available to facilitate sales on the world market.[12] The import levies and export

---

[6] Snyder, F. G., *Law of the Common Agricultural Policy*, at 143.

[7] COM(83)500, at 15.

[8] [1968] JO L148/13. For a discussion of the principles and objectives behind common organizations of the market in the agricultural sector, see, e.g., Snyder, F. G., *Law of the Common Agricultural Policy*, at 15–39; and Barents, R., *The Agricultural Law of the EC* (Kluwer, Deventer, 1994), at 73–171.

[9] The target price is 'that price which it is aimed to obtain for the aggregate of producers' milk sales, on the Community market and on external markets, during the milk year': Art.3(2). Following the implementation of the milk quota system, where a producer becomes liable to payment on excess production the rate of payment has been set by reference to this target price.

[10] The buying of Grana Padano and Parmigiano Reggiano cheeses has now been abolished (given their limited preservation time and the absence of disposal opportunities): Council Reg. (EC) 1880/94, [1994] OJ L197/21. The intervention prices for those two commodities were rescinded by Council Reg. (EC) 1881/94, [1994] OJ L197/23.

[11] The threshold prices are fixed 'so that, taking account of the protection required for the Community processing industry, the prices of imported milk products correspond to the level of the target price for milk': see e.g., Council Reg. (EC) 1882/94, [1994] OJ L197/24, Preamble.

[12] For discussion of the common organization of the market in milk and milk products, see, e.g., Snyder, F. G., *Law of the Common Agricultural Policy*, at 83–4; Usher, J. A., *Legal*

refunds, accordingly, ensure a strong degree of insulation against lower prices pertaining on the world market. Moreover, when taken together with the general benefits enjoyed by the agricultural sector (such as the modification of the competition rules),[13] this package of measures may be perceived as fulfilling a social function, not only bolstering the income of dairy farmers but arguably amounting to a 'welfare transfer'.[14]

The Community institutions had already enacted legislation within this framework to address the spiralling surplus in the dairy sector.[15] In 1977 a fixed rate and uniform levy, the 'co-responsibilty levy', had been imposed on milk delivered to dairies and on certain farm sales of milk products, allied with measures to expand the market.[16] However, while directly affecting producers' net receipts, the co-responsibility levy could not exceed 4 per cent of the target price for milk in the year in question;[17] and, at the time that decisions were being made to introduce milk quotas, it amounted to only some 2 per cent of that price.[18] This level of sanction proved inadequate to deter production — and, indeed, the co-reponsibility levy has now been abolished.[19] Subsequently, in 1982, a 'guarantee threshold' was implemented, fixed initially at the level of milk deliveries in the 1981 calendar year, plus 0.5 per cent. In the event of the guarantee threshold being exceeded, the Council was to adopt appropriate measures, which took the form of a cut in the milk price to offset the additional expenditure arising from the surplus.[20] The ensuing difficulties arising from the operation of this procedure may be illustrated by the Commission's estimate in 1983 that for the 1984/5 milk year the target price for milk price would have to be abated by some 12 per cent.[21]

Alongside the co-responsibility levy and the guarantee threshold, farmers had been encouraged, in return for premiums, to enter into undertakings not

Aspects of Agriculture in the European Community, at 72–3; and Court of Auditors, *Special Report No.4/93 on the Implementation of the Quota System intended to control Milk Production*, [1994] OJ C12/1, at 3.4–12.

[13] Art.42 of the EC Treaty.

[14] See, e.g., Conway, A. G., 'The Exchange Value of Milk Quotas in the Republic of Ireland and Some Future Issues for EC Quota Allocation', in Burrell, A. M. (ed.), *Milk Quotas in the European Community*, at 119–29.

[15] On the circumstances prevailing prior to the introduction of milk quotas, see, e.g., Harris, S., Swinbank, A., and Wilkinson, G., *The Food and Farm Policies of the European Community* (John Wiley, Chichester, 1983), at 323–41; and Neville-Rolfe, E., *The Politics of Agriculture in the European Community* (Policy Studies Institute, London, 1984), at 394–407.

[16] Council Reg. (EEC) 1079/77, [1977] OJ L131/6.   [17] Ibid., Art.2(3).

[18] Council Reg. (EEC) 1209/83, [1983] OJ L132/6.

[19] Council Reg. (EEC) 1029/93, [1993] OJ L108/4. Promotion of the market in milk and milk products had already been taken over by initiatives implemented under Council Reg. (EEC) 2073/92, [1992] OJ L215/67.

[20] Council Reg. (EEC) 1183/82, [1982] OJ L140/1; and Council Reg. (EEC) 1184/82, [1982] OJ L140/2.

[21] COM(83)500, at 17.

to market milk or milk products for specified periods, or to convert their dairy herds to meat production for specified periods.[22] These 'outgoers' schemes' had the twin aims of restoring balance in the milk market and promoting improvement in the production structures. Again, the impact was insufficient to curb the dairy surplus (although in the United Kingdom the total number of approved applications was just over 8,000, involving over 300,000 dairy cows and 1,350,000 tonnes of milk).[23] None the less, as shall be seen, the position of farmers who failed to secure reference quantities through compliance with such undertakings would prove one of the most controversial aspects of the implementation of the milk quota system.[24]

3. THE DECISION TO IMPOSE MILK QUOTAS

Faced by these difficulties in the operation of the guarantee threshold, the need for yet more fundamental change became apparent.[25] While the United Kingdom government argued for some time in favour of price cuts, the Council of Ministers finally agreed that the imposition of milk quotas would be preferable to either such price cuts or, indeed, a massive increase in the level of the co-responsibility levy.[26] The adoption of this drastic solution clearly reflected the critical state of imbalance that pervaded the dairy sector. Besides, in the view of the Commission, the need to carry out such reforms required more than technical adjustment of existing mechanisms: rather, it amounted to 'a political challenge'.[27] As stated in the implementing legislation itself, the increase in milk production was such that the disposal of the surplus was imposing financial burdens and market difficulties which were 'jeopardizing the very future of the common agricultural policy'.[28]

In this context budgetary considerations without doubt played a leading role. However, throughout the Community institutions also laid emphasis on the broader social and economic objectives, themselves enshrined in Article 39 of the EC Treaty — most notably, Article 39(1)(b), setting

[22] See, in particular, Council Reg. (EEC) 1078/77, [1977] OJ L131/1; and Comm. Reg. (EEC) 1307/77, [1977] OJ L150/24. Under Council Reg. (EEC) 1078/77 non-marketing undertakings were of a five year span and conversion schemes of a four year span. See also, e.g., Milk Marketing Board, *Five Years of Milk Quotas: a Progress Report* (Thames Ditton, Surrey, 1989), at 7.

[23] Ibid.   [24] For the litigation on this aspect, see Ch.2.

[25] For more detailed discussion of the political aspects of the introduction of milk quotas, see, e.g., Avery, G., 'The Common Agricultural Policy: a Turning Point?', (1984) 21 C.M.L. Rev. 481–504; and Petit, M. *et al.*, *Agricultural Policy Formation in the European Community: the Birth of Milk Quotas and CAP Reform, passim*.

[26] For a statement of the UK's position, see, e.g., *Hansard* (HC) Vol.63, Col.161.

[27] COM(83)500, at 1.

[28] Council Reg. (EEC) 856/84, [1984] OJ L90/10, Preamble. See also *Bulletin of the European Communities*, No.3, 1984, at 11 and 13.

out the need 'to ensure a fair standard of living for the agricultural community, in particular by increasing the individual earnings of persons engaged in agriculture'. Accordingly, one of the grounds for rejecting the alternative of abating the target price for milk by 12 per cent was that it would have 'grave and immediate effects on the revenues of producers';[29] and the implementing legislation affirmed that the adoption of milk quotas had the 'least drastic effect on the incomes of producers'.[30] Likewise, a reason for the rejection of any major increase in the level of the co-responsibility levy was that it would require differentiation to protect small producers; and the level of differentiation would be such as to run the risk of creating inequalities between Member States (and might even compromise the unity of the price mechanism).[31]

At the same time there was a strong awareness of the different priorities and different circumstances of the various Member States; and four illustrations of such factors highlight the potential frictions. First, there was — and still remains — a great divergence between the average size of dairy herd in the different Member States and in the average yield per dairy cow.[32] For example, in 1981 the average number of cows in a United Kingdom dairy herd was 56.1 (the highest number of any Member State); and the average annual yield per cow was 4,831 kg (an amount only exceeded in the Netherlands). By contrast, in Italy the respective figures were 6.4 and 3,394 kg.[33] Accordingly, it was perhaps not surprising that the United Kingdom resisted the proposals to weight the co-responsibility levy against larger farms.[34]

Secondly, the relative importance of the dairy sector varied considerably between Member States. For example, at the time of the introduction of milk quotas Ireland's dairy industry contributed directly or indirectly to some 9 per cent of the gross national product — a degree of reliance which justified special treatment.[35] By contrast, milk and milk products were of only fractional importance to the United Kingdom economy, although the overall figures masked considerable dependence in certain areas, most notably Northern Ireland.[36]

[29] COM(83)500, at 17.
[30] Council Reg. (EEC) 856/84, [1984] OJ L90/10, Preamble.   [31] COM(83)500, at 17.
[32] For discussions of farming structures in the dairy sector, see, e.g., Tollens, E., 'The Effects of Milk Quotas on Community Agriculture 1984–1987', in Burrell, A. M. (ed.), *Milk Quotas in the European Community*, at 183–92; and Milk Marketing Board, *Five Years of Milk Quotas: a Progress Report*, at 2–5. See also Dillen, M. and Tollens, E., *Milk Quotas; their Effects on Agriculture in the European Community* (Eurostat, 1990) ('Eurostat 1990'), Vol.1, at 137–242.
[33] European Commission, *The Agricultural Situation in the Community: 1983 Report*, at 291 and 365.
[34] See, e.g., *Hansard* (HL) Vol.454, Col.1757. A levy on intensive production (proposed COM(83)500, at 19–20) was also successfully resisted.
[35] Council Reg. (EEC) 856/84, [1984] OJ L90/10, Preamble.
[36] See, e.g., Conway, A. G., 'Milk Quota Review in relation to CAP Objectives, Single EC Market, and GATT Negotiations', Paper presented to the Centre for European Policy Studies

Thirdly, while the Community as a whole had reached self-sufficiency in the dairy sector as long ago as 1974, there was again considerable disparity between the situation in different Member States. Ireland, for example, produced 4,300 per cent of its requirements in whole-milk powder during the 1980 calendar year, but the United Kingdom and Germany produced only 429 and 144 per cent of their respective requirements in 1981. With regard to cheese, Ireland was 555 per cent self-sufficient in 1980; but the following year the United Kingdom failed to meet national demand, being only 71 per cent self-sufficient.[37] There is no doubt that the various Member States were very aware of this disparity. In the United Kingdom, notwithstanding major increases in production following entry into the Community,[38] it was widely perceived that there was no overall self-sufficiency in the dairy sector — with the concomitant fear that, on the introduction of a quota system based on historic production, the country would be locked into a position of permanent deficit.[39]

Fourthly, there were substantial differences in the structure of milk marketing in the various Member States. For example, while in Ireland there was a multiplicity of dairies, in the United Kingdom the market was dominated by five purchasers, the Milk Marketing Board of England and Wales, the Scottish Milk Marketing Board, the North of Scotland Milk Marketing Board, the Aberdeen and District Milk Marketing Board, and the Milk Marketing Board for Northern Ireland.[40] Prior to the revocation of the statutory milk marketing schemes in 1994 and 1995,[41] the boards were

---

Seminar, Brussels, 8 Mar. 1989, at 11. Consequently, Northern Ireland also received special treatment. See also Sheehy, S. J., 'Implications for an Exporting Member State of Alternative Reforms of the CAP Dairy Market', in Thomson, K. J. and Warren, R. M. (edd.), *Price and Market Policies in European Agriculture* (University of Newcastle-upon-Tyne, 1984), at 178–90.

[37] European Commission, *The Agricultural Situation in the Community: 1983 Report*, at 239–40.

[38] Ibid.

[39] It is very difficult to form an accurate overall picture in this regard. As seen, a Member State might enjoy a surplus in one part of the dairy sector, such as whole-milk powder, but a deficit in another, such as cheese. Further, the figures are complicated in the case of the UK by the guaranteed access arrangements for New Zealand products. At the time of the introduction of milk quotas the UK Government appeared sanguine as to the degree of national self-sufficiency (see, e.g., *Hansard* (HC) Vol.60, Cols.337–8); but later events have highlighted areas of deficit, in particular the inability to boost production to meet national demand following the abolition of the milk marketing schemes (see, e.g., Munir, A. E. 'Milk Marketing Upheaval', (1994) 138 Sol. J. 420–1). For an illustration of the level of imports at the time of the introduction of milk quotas, see, e.g., Written Answer, *Hansard* (HL) Vol.470, Cols.325–6. See also the *Third Report from the Agriculture Committee: Trading of Milk Quota*, (Session 1994–5, H.C. 512), at xi (UK quota only sufficient to fulfil 85% of domestic demand).

[40] For discussions of the milk marketing schemes, see, e.g., *Halsbury's Laws of England*, Vol.1(2) Agriculture, 4th Re-issue 1990, at 441–50; and Usher, J. A., *Legal Aspects of Agriculture in the European Community*, at 81–2.

[41] The statutory schemes governing the Milk Marketing Board of England and Wales, the Scottish Milk Marketing Board, the North of Scotland Milk Marketing Board, and the Aberdeen and District Milk Marketing Board were revoked on 1 Nov. 1994, pursuant to the

vested not only with special rights of purchase but also numerous other rights and functions, including the right to equalize prices paid to producers regardless of the intended use of the milk purchased.[42] Moreover, the position of the boards had received express sanction from the Community legislature (subject to certain conditions, imposed, *inter alia*, to ensure that the rights were applied consistently with the EC Treaty and the general principles of Community law).[43]

Against this background, and in the light of the Commission's clear proposals as set out in COM(83)500, it came as no great surprise when milk quotas were introduced on 2 April 1984. Indeed, there was some evidence of increased production in anticipation of their introduction, with a view to securing the highest possible reference quantity;[44] and, subsequently, in the course of litigation, the Commission expressly doubted whether such a turn of events was unforeseeable for the applicant, 'since the business circles concerned were quite well aware of the problem of increasing surpluses on the milk market'.[45]

4. THE IMPLEMENTING LEGISLATION

4.1. General

The introduction of milk quotas was effected by amendment to Council Regulation (EEC) 804/68, the regulation governing the common organization of the market in milk and milk products.[46] A new Article 5c was inserted by Council Regulation (EEC) 856/84,[47] which, together with Council Regulation (EEC) 857/84[48] and Commission Regulation (EEC) 1371/84,[49] provided the initial legislative framework. On inception the system was

---

Agriculture Act 1993. The statutory scheme governing the Milk Marketing Board for Northern Ireland was revoked on 1 Mar. 1995, pursuant to the Agriculture (N.I.) Order 1993, S.I.1993 No.2665 (N.I.10).

[42] The special rights of purchase have been held to extend to pasteurized milk (Case 372/88, *Milk Marketing Board of England and Wales* v. *Cricket St. Thomas Estate*, [1990] ECR I–1345, [1990] 2 CMLR 800); and to skimmed and semi-skimmed milk (Case 40/92, *Commission* v. *UK*, [1994] ECR I–989).

[43] Council Reg. (EEC) 1422/78, [1978] OJ L171/14; and Comm. Reg. (EEC) 1565/79, [1979] OJ L188/29.

[44] As Lord Mackie of Benshie said, reiterating the view of other peers, 'if you can see a quota system coming there is a strong practical reason for the policy of producing as much as you can in order to get a bigger quota when the blow falls:' *Hansard* (HL) Vol.454, Col.1777. Similar considerations would seem to have applied on the introduction of sheep and suckler cow premium quotas.

[45] Case 170/86, *Von Deetzen* v. *Hauptzollamt Hamburg-Jonas* ('*Von Deetzen I*'), [1988] ECR 2355, at 2364. See also *Hoddom & Kinmount Estates* v. *Secretary of State for Scotland*, [1992] 22 EG 118, where the foreseeability issue was considered in the context of a compensation claim following compulsory purchase.

[46] [1968] JO L148/13.   [47] [1984] OJ L90/10.

[48] [1984] OJ L90/13.   [49] [1984] OJ L132/11.

projected to last for five milk years; but, following extensions, it is now authorized to continue until the year 2000.[50]

4.2. Guaranteed total quantities for the Community and Member States

This implementing legislation distinguished between 'wholesale quota' in respect of milk and/or milk equivalent delivered to purchasers and 'direct sales quota' in respect of milk and/or milk equivalent sold directly for consumption, direct sales quota being by far the smaller category. Indeed, deliveries of wholesale quota accounted for 92 per cent of production in 1983.[51]

In the case of wholesale quota, a 'guaranteed total quantity' was laid down for the Community;[52] and from this each Member State received a national 'guaranteed total quantity', distributed as a rule on the basis of deliveries in the 1981 calendar year, plus 1 per cent. However, exceptions were made for Ireland and Italy, both of which were permitted to fix their guaranteed total quantities by reference to deliveries in 1983. The reason which justified such a derogation for Ireland was the importance of the dairy sector to the national economy, while in Italy milk production for 1981 was the lowest of the last ten years, the average yield per cow was lower than the Community average, and deliveries had increased in the period from 1981 to 1983 in line with a switch away from direct sales.[53] The

---

[50] See Council Reg. (EEC) 856/84, [1984] OJ L90/10 (the first period commencing, however, on 2 Apr. 1984 rather than 1 Apr. 1984); Council Reg. (EEC) 1109/88, [1988] OJ L110/27; Council Reg. (EEC) 816/92, [1992] OJ L86/83; and Council Reg. (EEC) 3950/92, [1992] OJ L405/1.

[51] Court of Auditors, *Special Report No.4/93 on the Implementation of the Quota System intended to control Milk Production*, [1994] OJ C12/1, at 2.13.

[52] Council Reg. (EEC) 856/84, [1984] OJ L90/10, Preamble. According to the Preamble the guaranteed total quantity for the Community as a whole was fixed at 97.2 million tonnes of milk or milk equivalent, which corresponded with the guarantee threshold laid down by the Council in 1983. However, it also stated that, by way of 'transitional relief', this figure was to be increased to 98.2 million tonnes for the first year of the system. When the Community reserve and transfers from direct sales to wholesale quota were further taken into account, the Court of Auditors could calculate that the true total for the 1984/5 milk year amounted to 99.9 million tonnes: *Special Report No.2/87 on the Quota/Additional Levy System in the Milk Sector*, [1987] OJ C266/1, at 3.10. See also COM(86)645, at Annex V.

[53] Council Reg. (EEC) 856/84, [1984] OJ L90/10, Preamble. See also COM(86)645, at Annex V. Following the accession of new Member States, the guaranteed total quantity for the Community has been correspondingly increased and distributed to those Member States. For dispute as to the terms of Spain's accession, see Case 203/86, *Spain* v. *Council*, [1988] ECR 4563. For implementation of Portugal's guaranteed total quantity and for increase in Germany's guaranteed total quantity to take into account unification with the territory of the former German Democratic Republic, see Council Reg. (EEC) 3641/90, [1990] OJ L362/5. For the provisions applicable to Austria, Finland, and Sweden, see the 1994 Act of Accession. In the case of all Member States joining after the implementation of the milk quota system, exceptions were permitted from the general rule which fixed the guaranteed total quantity by reference to 1981 deliveries. In this context it may be noted that, following 'special examina-

United Kingdom, in accordance with the general rule, received an initial guaranteed total quantity of 15,487,000 tonnes.[54]

Provision was also made for a Community reserve to combat structural difficulties. For the first year of the system the reserve was eventually fixed at 393,000 tonnes.[55] From this the United Kingdom received 65,000 tonnes in respect of the region of Northern Ireland, to be added to the 15,487,000 tonnes already mentioned.[56] Such provision, together with the exceptions in favour of Ireland and Italy already mentioned, illustrate from inception the willingness of the Community institutions to accommodate the differing economic and social factors prevailing. However, there was also concern that milk quotas should not ossify structural development; and Council Regulation (EEC) 856/84 itself stated that the objective was 'to curb the increase in milk production while at the same time permitting the structural developments and adjustments required, having regard to the diversity of the situations among individual Member States, regions and collection areas in the Community'. Likewise, the promotion of agricultural efficiency found expression the following year with the enactment of Council Regulation (EEC) 797/85.[57]

In the case of direct sales quota, each Member State was again subjected to a national ceiling.[58] Following amendment, the United Kingdom received a total of 398,000 tonnes for the 1984/5 milk year — a mere fraction of the guaranteed total quantity for wholesale quota.[59]

From the 1984/5 milk year it was a cornerstone of the system that a superlevy became payable if a Member State exceeded its guaranteed total quantity for wholesale and/or direct sales quota.[60] That having been said, there is a good argument that overall effectiveness was hampered by the setting of the initial guaranteed total quantities at too high a level, with the result that, notwithstanding reductions in those quantities over the years,

tion' of the implementation of the milk quota system in Greece, Italy, and Spain, substantial increases were established for all three Member States: Council Reg. (EC) 1883/94, [1994] OJ L197/25; and Council Reg. (EC) 1552/95, [1995] OJ L148/43.

[54] Council Reg. (EEC) 1557/84, [1984] OJ L150/6.

[55] Council Reg. (EEC) 1298/85, [1985] OJ L137/5.

[56] The first allocation was made under Comm. Reg. (EEC) 1371/84, [1984] OJ L132/11. The greatest beneficiary was Ireland, which eventually received 303,000 tonnes for the 1984/5 milk year. For the special position of Northern Ireland, see, e.g., Kirke, A. W., 'The Influence of Milk Supply Quotas on Dairy Farm Performance in Northern Ireland', in Burrell, A. M. (ed.), *Milk Quotas in the European Community*, at 30–45.

[57] [1985] OJ L93/1.

[58] Council Reg. (EEC) 857/84, [1984] OJ L90/13, Art.6(2) and Annex. This did not specifically refer to a 'guaranteed total quantity'; but the expression 'guaranteed total quantity for direct sales quota' has commonly been employed.

[59] Council Reg. (EEC) 1557/84, [1984] OJ L150/6. For the guaranteed total quantities for direct sales quota for all Member States, see COM(86)645, at Annex VII.

[60] Since the milk quota system imposed a potential levy over and above the co-responsibility levy, the term 'superlevy' or 'additional levy' was adopted. Following the abolition of the co-responsibility levy, it is perhaps now less appropriate.

serious problems have been encountered in balancing supply with demand.[61]

4.3. Allocations within the Member State: general

For the purposes of allocations within Member States, different provisions applied in respect of wholesale and direct sales quota. In the case of wholesale quota, the original legislation entitled Member States to choose one of two methods for the operation of the system within each region of its territory. For these purposes a 'region' was defined as meaning 'all or part of the territory of a Member State having geographical unity and in which the natural conditions, the structures of production and the average yields of the herds are comparable'.[62] Initially, the United Kingdom operated very much on a regional basis, with the power to determine regions being conferred upon the Minister of Agriculture;[63] but currently the whole territory is treated as one region, with the exception of the 'Scottish Islands area'.[64]

Under the first method, Formula A, each individual producer received a reference quantity in respect of his holding and paid the superlevy on his own excess production. On the introduction of the milk quota system the rate of levy was fixed at 75 per cent of the target price for milk. Under the second method, Formula B, purchasers received reference quantities and paid the superlevy on deliveries made to them in excess of those reference quantities. The burden was than passed on to individual producers in proportion to their contribution to the excess. Where Formula B was applied, the rate of superlevy was initially fixed at 100 per cent of the target price for milk.[65] This higher rate was implemented on the basis that under Formula B producers were unlikely to be subject to superlevy on all their excess

[61] See, e.g., Court of Auditors, *Special Report No.2/87 on the Quota/Additional Levy System in the Milk Sector*, [1987] OJ C266/1, at 3.10.

[62] Council Reg. (EEC) 857/84, [1984] OJ L90/13, Art.1(2).

[63] The Dairy Produce Quotas Regs. 1984, S.I.1984 No.1047, reg.5. See also Wood, D., *et al.*, *Milk Quotas: Law and Practice*, at 8.

[64] I.e., any one of (a) the islands of Shetland; (b) the islands of Orkney; (c) the islands of Islay, Jura, Gigha, Arran, Bute, Great Cumbrae and Little Cumbrae and the Kintyre peninsula south of Tarbert; or (d) the islands in the Outer Hebrides and the Inner Hebrides other than those listed in (c): the Dairy Produce Quotas Regs. 1994, S.I.1994 No.672, reg.2(1). A Member State may treat the whole of its territory as a single region, even if the territory is not a geographical unit in which the natural conditions, structures of production, and average herd yields are comparable, unless such a decision is manifestly unsuited to the structures of the Member State in question: Joined Cases 267–285/88, *Wuidart* v. *Laiterie coopérative eupenoise*, [1990] ECR I–435. Accordingly, the Member States retain a broad discretion in appraising the requisite criteria.

[65] However, subsequently the rate of levy under Formula A was raised to 100%, as under Formula B: Council Reg. (EEC) 1305/85, [1985] OJ L137/12 (in the case of producer groups and associations — as referred to in Art.12(c) of Council Reg. (EEC) 857/84); and Council Reg. (EEC) 774/87, [1987] OJ L78/3 (in the case of all other Formula A producers).

production: in all probability other producers supplying the same purchaser would deliver less than their individual reference quantities, giving rise to the opportunity for 'offsetting' at the level of the purchaser.[66]

The choice of formula was to comply with one or more of three criteria: first, administrative viability; secondly, the need to facilitate structural change and adaptation; and, thirdly, regional development requirements, in which regard one consideration was the need to avoid desertification of certain areas.[67] The United Kingdom opted for Formula B (with the exception of the Northern Ireland region and the Isles of Scilly, which for the 1984/5 milk year only were subject to Formula A). The decision may be regarded as consistent with the position of the five milk marketing boards as both dominant purchasers and administrators of the national milk market. Not least, with so many producers making deliveries to so few purchasers, there was great scope for offsetting. Formula B was also adopted by the majority of Member States, Denmark deriving maximum advantage through a central milk collecting agency.[68]

In the case of direct sales quota, producers received a reference quantity and paid the superlevy on their individual excess sales. At the commencement of the milk quota system, the rate of superlevy was fixed at 75 per cent of the target price for milk. Once again the reduced rate was applicable on the basis that there was more individual responsibility for over-production.[69]

Since this régime formed an intervention measure, any revenue generated (whether in respect of wholesale or direct sales quota) was allocated to the financing of expenditure in the milk and milk products sector.[70] Member States were soon authorized to employ such sums to grant compensation under outgoers' schemes.[71] While initially the collection of levy in respect of wholesale quota was to be carried out by means of quarterly payments on

---

[66] For offsetting, see further Ch.3, 1.2.1. The European Court confirmed that no superlevy would be payable under Formula B unless the purchaser's reference quantity was exceeded, notwithstanding that certain producers supplying that purchaser had exceeded their individual reference quantities: Case 61/87, *Thevenot* v. *Centrale Laitière de Franche-Comté*, [1988] ECR 2375, [1989] 3 CMLR 389.

[67] Council Reg. (EEC) 857/84, [1984] OJ L90/13, Art.1(2).

[68] For the choice of formula effective in the various Member States during the 1984/5 milk year, see COM(86)645, at 6; and Eurostat 1990. Vol.1, at Table 1. In the case of Denmark, see Walter-Jorgensen, A., 'The Impact of Milk Quotas in Denmark', in Burrell, A. M. (ed.), *Milk Quotas in the European Community*, at 21–9. For the factors infuencing this choice, see, e.g., Harvey, D. R., *Milk Quotas: Freedom or Serfdom?*, at 27–8; and Gehrke, H., *The Implementation of the EC Milk Quota Regulations in British, French and German Law*, at 35–8 and 62–5.

[69] Council Reg. (EEC) 857/84, [1984] OJ L90/13, Art.1(1). [Under the Commission's proposals, as set out in COM(83)500, only the Formula B method of collection was envisaged (at 19). However, Germany argued strongly in favour of providing for Formula A (the alternative which it subsequently adopted).]

[70] Council Reg. (EEC) 856/84, [1984] OJ L90/10.

[71] Council Reg. (EEC) 1298/85, [1985] OJ L137/5, and Council Reg. (EEC) 1305/85, [1985] OJ L137/12.

account,[72] administrative difficulties resulted in the substitution of collection on an annual basis, accompanied by half-yearly provisional statements.[73] The collection of levy in respect of direct sales quota was to be carried out, at the latest, within four months of the end of the milk year in question.[74] The Intervention Board for Agricultural Produce, constituted under the European Communities Act 1972, was the body in the United Kingdom entrusted with the task of collecting any superlevy due. However, it discharged these functions to the five milk marketing boards.[75] For the purposes of collecting the superlevy and the purposes of administration generally, the Minister was required to maintain a register of direct sales producers and a register of wholesale producers. However, again the function was discharged to the milk marketing boards.[76]

A critical feature from the inception of the milk quota system was that in all cases it became necessary to determine a reference quantity for each producer (even if under Formula B this was for the purpose of ascertaining the proportionate share of the producer's contribution to his purchaser's superlevy). This feature remains critical to date, notwithstanding the numerous amendments and repeals effected.[77]

4.4. Allocations within the Member State: producers and purchasers

The legislation governing the determination of reference quantities for producers and purchasers was complex, reflecting the twin objectives of curbing milk production and permitting structural developments and adjustments. In the case of wholesale quota the provisions were materially different depending upon whether a Member State adopted Formula A or Formula B. If it adopted Formula A, each producer in principle received a reference quantity equal to the quantity of milk or milk equivalent delivered in the 1981 calendar year. This contrasted with the position under Formula B, where the purchaser received in principle a reference

---

[72] Council Reg. (EEC) 857/84, [1984] OJ L90/13, Art.9.

[73] Council Reg. (EEC) 1305/85, [1985] OJ L137/12.

[74] Comm. Reg. (EEC) 1371/84, [1984] OJ L132/11, Art.13(2) and (3); and see also the Dairy Produce Quotas Regs. 1984, S.I.1984 No.1047, reg.11.

[75] Authority to this effect was granted, under the initial legislation, by ibid., reg.10. However, following the reforms under the Agriculture Act 1993 and, in the case of Northern Ireland, the Agriculture (N.I.) Order 1993, S.I.1993 No.2665 (N.I.10), the Intervention Board now has immediate responsibility for collecting payment: the Dairy Produce Quotas Regs. 1994, S.I.1994 No.672, reg.23. On this aspect, see Ch.3, 2.2.4.

[76] The Dairy Produce Quotas Regs. 1984, S.I.1984 No.1047, Sch.1, para.14 (in respect of direct sales quota) and Sch.2, para.14 (in respect of wholesale quota). Authority to discharge this function on behalf of the Minister was granted by ibid., reg.13. Following the reforms, the Intervention Board has also taken over this task: the Dairy Produce Quotas Regs. 1994, S.I.1994 No.672, reg.25, as amended by the Dairy Produce Quotas (Amendment) Regs. 1994, S.I.1994 No.2448.

[77] See, e.g., the dicta of Chadwick J in *Faulks* v. *Faulks*, [1992] 15 EG 82, at 85.

quantity equal to the quantity of milk or milk equivalent purchased in the 1981 calendar year, plus 1 per cent.[78]

These provisions were supplemented by derogations to ensure the requisite degree of flexibility. First, the general rule governing the 'reference' or 'base' year was relaxed.[79] Under both Formula A and Formula B Member States could choose to make their overall allocation of reference quantities on the basis of deliveries or purchases in the 1982 or 1983 calendar years (rather than deliveries or purchases in the 1981 calendar year). Should this discretion be exercised, an appropriate percentage weighting was to be applied so that the sum of allocations would not exceed the national quaranteed total quantity. However, to enhance this flexibility, Member States could vary the percentage weighting on the basis of the level of deliveries of certain categories of persons liable for the superlevy, on the basis of the trend in deliveries in certain regions between 1981 and 1983, or on the basis of the trend of deliveries of certain categories of persons liable during that period.[80]

Further, individual producers affected by certain 'exceptional events' occurring before or during their Member State's reference year were entitled to opt for a different reference year within the 1981–3 period. The six exceptional events which could trigger such an application were: first, a serious natural disaster which affected the producer's farm to a substantial extent; secondly, the accidental destruction of the producer's fodder resources or buildings used for dairy livestock; thirdly, an epizootic which affected all or part of the milk herd (i.e., a temporarily prevalent disease); fourthly, compulsory appropriation of a considerable part of the utilizable agricultural area of the producer's holding, resulting in a temporary reduction of the fodder area of the holding;[81] fifthly, occupational incapacity of long duration where the producer farmed the holding himself; and, sixthly, the theft or accidental loss of all or part of the dairy herd, where this had a significant effect on the milk production of the holding.[82]

---

[78] Council Reg. (EEC) 857/84, [1984] OJ L90/13, Art.2(1). In determining individual reference quantities, a Member State could justifiably refuse to take into account milk produced on a holding in another Member State: Case 351/92, *Graff* v. *Hauptzollamt Köln-Rheinau*, [1994] ECR I–3361, [1995] 3 CMLR 152.

[79] The reference or base year in this context (i.e., for the purposes of allocations within a Member State) is to be distinguished from the reference or base year for the purposes of determining a Member State's guaranteed total quantity.

[80] Council Reg. (EEC) 857/84, [1984] OJ L90/13, Art.2(2). See also Comm. Reg. (EEC) 1371/84, [1984] OJ L132/11, Art.2. For subsequent amendment of these provisions, see Council Reg. (EEC) 1343/86, [1986] OJ L119/34; and Council Reg. (EEC) 1911/86, [1986] OJ L165/6.

[81] See Case 285/89, *Van der Laan-Velzeboer* v. *Minister for Agriculture and Fisheries*, [1990] ECR I–4727: the Court held that this exceptional event extended to circumstances where a producer entered into an agreement in order to avoid unilateral imposition of an obligation to tolerate public works.

[82] Council Reg. (EEC) 857/84, [1984] OJ L90/13, Art.3(3), as expanded by Comm. Reg. (EEC) 1371/84, [1984] OJ L132/11, Art.3.

A second derogation related to producers who had adopted milk production development plans under Directive 72/159/EEC[83] lodged before 1 March 1984. In effect, this derogation was designed to address the difficult circumstances faced by farmers who had committed themselves to such development plans but who, owing to the introduction of quotas, were effectively precluded from reaping the fruits of their investments.[84] It was provided that such producers 'may obtain' a special reference quantity 'according to their Member State's decision'.[85] There has been uncertainty as to the true effect of this provision and, in particular, whether it imposes an obligation to grant a special reference quantity to this category of producers or merely confers a discretion upon Member States. In the course of its judgment in *Cornée* v. *Coopérative agricole laitière de Loudéac and Laiterie coopérative du Trieux* the European Court referred to the derogation's discretionary nature;[86] but the decision of the European Court is awaited in a later case where the applicant in terms claims that Member States are obliged to allocate a special reference quantity — and, should it be otherwise, that the regulation is to that extent invalid.[87]

A third derogation, clearly at the discretion of the Member State, permitted the allocation of a 'specific reference quantity' to young farmers setting up after 31 December 1980.[88]

Finally, Member States could make allocations to producers in order to complete the restructuring of milk production at national level, or regional level, or at the level of the collecting areas.[89] Under this provision an 'additional reference quantity' could be granted to producers realizing a development plan approved after the implementation of the milk quota system under Directive 72/159/EEC,[90] subject to the plan meeting certain criteria.[91] Further, an 'additional reference quantity' could be granted to producers undertaking farming as their main occupation. In tandem with

---

[83] [1972] JO L96/1.

[84] Council Reg. (EEC) 857/84, [1984] OJ L90/13, Art.3(1).

[85] If the plan was still being implemented, the special reference quantity was to take into account the milk and milk product quantities provided for in the development plan. If the plan had been implemented after 1 Jan. 1981, the special reference quantity was to take into account the milk and milk product quantities delivered in the year that the plan was completed. Further, investments carried out without a development plan could be taken into account if the Member State had sufficient information.

[86] Joined Cases 196–198/88, [1989] ECR 2309.

[87] Case 63/93, *Duff* v. *Minister for Agriculture and Food.* See also [1993] 2 CMLR 969, [1994] IJEL 247 (Irish High Court); [1993] 2 CMLR 969 (Irish Supreme Court). The case also considers the effect of approval of the development plans by the competent authority.

[88] Council Reg. (EEC) 857/84, [1984] OJ L90/13, Art.3(2).

[89] Ibid., Art.4(1)(b) and (c).

[90] [1972] JO L96/1.

[91] The criteria were set out in Council Reg. (EEC) 1946/81, [1981] OJ L197/32, Art.1(2).

discretionary allocations of quota, restructuring could also be pursued by the implementation of outgoers' schemes, Member States being authorized to grant compensation to producers undertaking to discontinue milk production definitively.[92]

A national reserve was to be constituted for the purposes of implementing the various derogations; and, in particular, reference quantities freed under the outgoers' schemes were, as necessary, to be added to this reserve for the purposes of reallocation.[93] However, while these provisions did indeed promote flexibility, any exercise of the derogations could only be made within the Member State's guaranteed total quantity, a stipulation which imposed severe restraint.[94]

In the case of direct sales quota, each producer was to receive a reference quantity corresponding to his direct sales during the 1981 calendar year, plus 1 per cent.[95] The legislation as first enacted did not extend to direct sales quota the discretion conferred on Member States to select either 1982 or 1983 as an alternative reference year for making such allocations. That having been said, the disparity was soon rectified by Council Regulation (EEC) 590/85;[96] and in all other respects the same flexibility was available *ab initio*.[97]

The United Kingdom implemented some, but not all, of these derogations. In line with most other Member States, the option was exercised to fix allocations of wholesale quota by reference to the 1983 calendar year, thus permitting the allocations to conform with production trends shortly before the imposition of milk quotas.[98] As a consequence, there having been a substantial increase in production between 1981 and 1983, it proved necessary to provide for, in general, a 9 per cent reduction in individual producers' reference quantities so as not to exceed the national guaranteed quantity.[99] Further, in accordance with their entitlement under the Community legislation, it was provided that direct sales and wholesale

---

[92] Council Reg. (EEC) 857/84, [1984] OJ L90/13, Art.4(1)(a).

[93] Ibid., Arts.4(2) and 5.   [94] Ibid., Art.5.

[95] Ibid., Art.6(1). The calculation was to be made on the basis of direct sales of milk and milk products from the producer's own herd; no account could be taken of milk or milk products bought in and sold on: Case 174/88, *R.* v. *Dairy Produce Quota Tribunal for England and Wales*, ex p. *Hall & Sons (Dairy Farmers) Ltd.*, [1990] ECR I–2237.

[96] [1985] OJ L68/1. See also Comm. Reg. (EEC) 1043/85, [1985] OJ L112/18.

[97] Council Reg. (EEC) 857/84, [1984] OJ L90/13, Art.6(3).

[98] For the choice of reference year by all Member States, see COM(86)645, at 6; and Eurostat 1990, Vol.1, at Table 1 (all but Greece and Luxembourg adopting the 1983 calendar year). There is immense variety in the detailed calculation of reference quantities throughout the Member States, with preferential treatment frequently being accorded to small producers.

[99] The Dairy Produce Quotas Regs. 1984, S.I.1984 No.1047, Sch.2, para.4(c). The UK also exercised the option to vary the level of reduction: e.g., in the case of 'Scottish area B' (Kintyre, south of Tarbet, and the islands of Arran, Bute, Coll, Gigha, Great Cumbrae, Islay, Little Cumbrae, and Orkney) the reduction was limited to 5.8135%. See also *First Report from the Agriculture Committee: the Implementation of Dairy Quotas* (Session 1984–5, H.C.14).

producers could apply for one of the alternative reference years if affected by an exceptional event. This application was termed a 'base year revision claim'.[100] However, where the applicant relied on a serious natural disaster caused by weather, he was required to show that in consequence milk production was reduced in the base year by no less than 15 per cent.[101]

The United Kingdom also exercised its discretion to make allocations to farmers who had adopted milk production development plans within Article 3(1) of Council Regulation (EEC) 857/84.[102] Applications under this head were termed 'development claims'.[103] Similar provisions implemented on the introduction of sheep and suckler cow premium quotas have proved controversial, the United Kingdom legislation being successfully challenged on the basis that it imposed eligibility criteria over and above those stipulated in the Community rules.[104]

Further, considerable use was made of the derogations designed to permit restructuring. The 'exceptional hardship claim' was introduced under the authority of Article 4(1)(c) of Council Regulation (EEC) 857/84 (which permitted Member States to grant additional reference quantities to producers undertaking farming as their main occupation).[105] To qualify for such an award, the claimant was obliged to show, *inter alia*, that, prior to the introduction of milk quotas on 2 April 1984, he had entered into, or become obliged to enter into, a transaction or had made an arrangement as a result of which his direct sales or wholesale reference quantity was substantially less than it would otherwise have been, or the reasonably expected outcome of which was a level of direct sales or deliveries which otherwise was not covered, or a substantial part of which was not covered, by allocations under the regulations. The same Article 4(1)(c) subsequently supplied the authority for further allocations directed to the specific needs of certain categories of producer. In this context may be highlighted the 'small producer provision' which was made available, as a rule, to producers the aggregate of whose direct sales and wholesale quota was less than 200,000 litres at a date determined in accordance with the regulations.[106]

Adjudication of base year revision claims, development claims and

[100] See the Dairy Produce Quotas Regs. 1984, S.I.1984 No.1047, Sch.1, paras.8–13 (in respect of direct sales quota); and Sch.2, paras.8–13 (in respect of wholesale quota).

[101] The Dairy Produce Quotas (Definition of Base Year Revision Claims) Regs. 1984, S.I.1984 No.1048.

[102] [1984] OJ L90/13.

[103] See the Dairy Produce Quotas Regs. 1984, S.I.1984 No.1047, Sch.1, paras.8–13 (in respect of direct sales quota); and Sch.2, paras.8–13 (in respect of wholesale quota).

[104] The Sheep Annual Premium and Suckler Cow Premium Quotas Regs. 1993, S.I.1993 No.1626, as amended by the Sheep Annual Premium and Suckler Cow Premium Quotas (Amendment) Regs. 1993, S.I.1993 No.3036; and *R.* v. *Ministry of Agriculture, Fisheries and Food*, ex p. *National Farmers Union*, [1995] 3 CMLR 116.

[105] The Dairy Produce Quotas Regs. 1984, S.I.1984 No.1047, Sch.1, para.17 (in respect of direct sales quota); and Sch.2, para.17 (in respect of wholesale quota).

[106] The small producer provision was introduced by the Dairy Produce Quotas (Amendment) Regs. 1985, S.I.1985 No.509, regs.8 and 16 and Sch.1. For the Government's determina-

exceptional hardship claims was undertaken by a combination of the Minister, 'local panels', and Dairy Produce Quota Tribunals.[107] The administrative burden was great, and by September 1984 local panels had considered 23,189 applications in respect of base year revision claims and development claims.[108] Nonetheless, by 1986 much of the task had been completed and it was not felt appropriate to continue the existence of local panels.[109] By contrast, Dairy Produce Quota Tribunals remain in place to date.[110]

Between them these further allocations did much to mitigate the difficulties experienced following the introduction of the milk quota system. Their effect may be judged by the size of the awards. For example, in the 1984/5 milk year 69 million litres of wholesale quota had been allocated consequent upon base year revision claims; 347 million litres consequent upon development claims; and 53 million litres consequent upon exceptional hardship claims. Further, under the small producer provision alone some 14,650 producers received approximately 163 million litres of wholesale quota in the 1985/6 milk year.[111] That having been said, the national reserve proved insufficient to meet demand. While base year revision claims were met in full, producers only received 75 per cent of exceptional hardship awards in the 1984/5 milk year, the balance being made up the following milk year.[112] Indeed, development claim allocations were reduced to approximately 65 per cent of the amount awarded, the balance being effectively made up following the provision of further reference quantities to Member States under the 'Nallet Package'.[113]

At the same time an outgoers' scheme was put into effect, initially on a non-statutory basis.[114] Directed at small producers (i.e., those producing less

tion to support this category, see *Hansard* (HC) Vol.64, Col.442. The Dairy Produce Quotas (Amendment) Regs. 1985 also made available the remote areas wholesale provision and the remote areas direct sales provision: regs.8 and 16 and Schs.4 and 5. Yet further categories were added later.

[107] For the procedure as originally enacted, see the Dairy Produce Quotas Regs. 1984, S.I.1984 No.1047, Sch.1, paras.8–13 and 17 (in respect of direct sales quota); and Sch.2, paras.8–13 and 17 (in respect of wholesale quota). It may be noted that the Dairy Produce Quota Tribunals, rather than local panels, were responsible for considering exceptional hardship claims. For the initial constitution of the Dairy Produce Quota Tribunals, see ibid., reg.6 and Sch.5. In Scotland there were no local panels.

[108] Milk Marketing Board, *Five Years of Milk Quotas: a Progress Report*, at 26. Before the end of 1984 the maximum number of members of tribunals for England and Wales and for Northern Ireland had increased from 12 to 90; and the maximum number of members of local panels had increased from 7 to 15: the Dairy Produce Quotas (Amendment) Regs. 1984, S.I.1984 No.1538; and the Dairy Produce Quotas (Amendment) (No.2) Regs. 1984, S.I.1984 No.1787.

[109] The Dairy Produce Quotas Regs. 1986, S.I.1986 No.470, reg.37(2).

[110] The Dairy Produce Quotas Regs. 1994 S.I.1994 No.672, reg.35 and Sch.6.

[111] Milk Marketing Board, *Five Years of Milk Quotas: a Progress Report*, at Table 4. Indeed, allocations of 'secondary quota' were said to amount to about 5.5% of total quota.

[112] Ibid., at 27.   [113] For which, see Ch.2, 3.

[114] Statutory authority for payments made was subsequently provided by the Milk (Cessation of Production) Act 1985. Schemes were thereafter to be made by statutory instrument.

than 200,000 litres per annum), the scheme provided them with the opportunity to leave the dairy sector altogether at so delicate a juncture; and, in accordance with the Community legislation, the reference quantities freed to the national reserve fuelled the further allocations.[115] By contrast, the United Kingdom did not enact the derogation in favour of young farmers.

Three aspects of this regulatory framework may be emphasized. First, while the numerous derogations ensured that the potential rigidity of the quota system was significantly relaxed, they did not cover all categories of producers materially affected. The European Court was soon to establish that they did not address the situation of those farmers who failed to secure a reference quantity through compliance with non-marketing or conversion undertakings.[116]

Secondly, in order to meet situations of hardship and maintain scope for structural developments and adjustments, it proved necessary to enact legislation of great complexity, giving rise to an equally complex administrative procedure.[117] Such complexity at once attracted adverse comment: for example, in the House of Commons debate on the Dairy Produce Quotas Regulations 1984 it was stated that 'the least uncomplimentary thing that can be said about them is that they will provide a feast for the lawyers and paradise for them in terms of interpretation'.[118] With the accretion of numerous amendments over the years, the Court of Auditors could come to see this characteristic of the milk quota system as one significantly reducing its effectiveness.[119]

Finally, it is possible to identify social and economic forces shaping the structure of the Community legislation. Certain instances have already been noted. For example, quotas were preferred to substantial increases in the co-responsibility levy on the ground that they would have less impact on the incomes of producers; and Ireland received the greatest allocation from the Community reserve in light of the dairy industry's importance to the national economy. None the less, as also noted, the Community institutions did not wish for the imposition of quotas to obviate structural development; and the continued migration of milk production to larger enterprises would appear to have been accepted. Moreover, certain measures would appear to have met both social and structural objec-

---

[115] Eligibility was extended in Nov. 1984 to producers with production of less than 275,000 litres p.a.; and by June 1985 some 1,670 producers had released over 270 millon litres of quota: Milk Marketing Board, *Five Years of Milk Quotas: a Progress Report*, at 26.

[116] For which, see Ch.2, 4.

[117] For express recognition of the difficulties involved, see Council Reg. (EEC) 856/84, [1984] OJ L90/10, Preamble.

[118] *Hansard* (HC) Vol.63, Col.172 (Mr Robert Hughes).

[119] *Special Report No.4/93 on the Implementation of the Quota System intended to control Milk Production*, [1994] OJ C12/1, at 4.9–11.

tives, notably the option to award specific reference quantities to young farmers.

There is an argument that in implementing quotas the United Kingdom accorded higher priority to aspects of structural development than did the Community institutions or, indeed, many of the other Member States. Thus, the Government initially supported a substantial increase in the co-reponsibility levy as opposed to the introduction of quotas;[120] and, despite vociferous appeals, no provision has been enacted with the express purpose of assisting young farmers — which may be contrasted with the position in, for example, France and Germany.[121] That having been said, both the initial and subsequent outgoers' schemes were expressly intended to provide hard-pressed small producers with an exit from the industry, while those who remained might benefit from the small producer provision. Moreover, Northern Ireland received 65,000 tonnes from the Community reserve; and the Scottish Islands area received favourable treatment in the making of initial allocations to producers. Accordingly, there is good evidence of pursuit of social objectives; but the overall picture is consistent with emphasis upon the continued development of, relative to most Member States, a substantially developed industry.

5. THE DEFINITION OF 'PRODUCER' AND 'HOLDING'

Before considering legal challenges to the implementing legislation, the concept of the 'producer' and the 'holding' may be explored in greater detail, both having remained central to the operation of the milk quota system.[122]

The definition of 'producer' was originally contained in Article 12(c) of Council Regulation (EEC) 857/84.[123] For the purpose of the regulation, a

---

[120] Government policy is well set out in the *First Report from the Agriculture Committee: the Implementation of Dairy Quotas* (Session 1984–5, H.C.14).

[121] See, e.g., Hairy, D. and Prost, M., 'Milk Quotas in France: Problems of Management', in Burrell, A. M. (ed.), *Milk Quotas in the European Community,* at 7–20; and Gehrke, H., *The Implementation of the EC Milk Quota Regulations in British, French and German Law*, at 81–4, 89, and 111.

[122] As shall be seen, there is a close relationship between 'producers' and 'holdings'. Thus, the original UK legislation provided that the appropriate registers should include details of the producer and identification of his holding: the Dairy Produce Quotas Regs. 1984, S.I.1984 No.1047, Sch.1, para.14 (in respect of direct sales quota); and Sch.2, para.14 (in respect of wholesale quota). However, the Community provision required only the address of the producer: Comm. Reg. (EEC) 1371/84, [1984] OJ L132/11, Art.11; and this lesser requirement is now stipulated in the current UK provisions: the Dairy Produce Quotas Regs. 1994, S.I.1994 No.672, reg.25. While such simplifies the process of registration, in certain circumstances it none the less becomes critical to identify the producer's holding, most notably in the event of quota transfers. For recent affirmation of the close relationship between 'producers' and 'holdings', see *Cottle* v. *Coldicott*, [1995] SpC 40.

[123] [1984] OJ L90/13.

producer was: 'a natural or legal person or group of natural or legal persons farming a holding located within the geographical territory of the Community: — selling milk or other milk products directly to the consumer, and/or — supplying the purchaser'. The definition of 'holding', contained in Article 12(d), was: 'all the production units operated by the producer and located within the geographical territory of the Community'.[124] 'Production units' were not defined. Notwithstanding the numerous and significant amendments to the milk quota system adopted over the years, these two definitions have remained remarkably unchanged.[125] Indeed, it is arguable that no material alteration was effected until 1993, when 'for reasons of control' the expression 'holding' was limited to the production units contained within a single Member State.[126] More specifically, this had the effect of preventing quota from crossing national frontiers.

Turning first to the definition of 'producer', on the face of the Community legislation this embraces both farmers making direct sales and farmers making wholesale deliveries. It also embraces tenants as well as freehold owners — as assumed throughout in United Kingdom legislation (such as the Agriculture Act 1986) and expressly confirmed by the European Court in *Ballmann* v. *Hauptzollamt Osnabrück*.[127] Further, with regard to business structures, it would seem to cover not only the farmer trading on his own account, but also partnerships and companies. However, two areas of potential difficulty may be considered. The first concerns the legal status of the producer and, in particular, problems which may arise when the trading entity is neither the freehold owner nor the tenant. The second concerns the extent to which a producer must be actively engaged in milk production. A third area of potential difficulty, the effect of alterations to business structures, will be considered later in the context of quota transfers.

Turning to the first of these aspects, it would appear in the case of partnerships that the general practice has been to register the reference

[124] For the definitions in the original UK implementing legislation, see the Dairy Produce Quotas Regs. 1984, S.I.1984 No.1047, reg.2(1): in the case of 'producer', the Community definition was adopted without qualification, but, in the case of 'holding', there were minor qualifications.

[125] In the reforming legislation, for which see Ch. 3, 'producer' was defined as 'a natural or legal person or a group of natural or legal persons farming a holding within the geographical territory of the Community: — selling milk or other milk products directly to the consumer, — and/or supplying the purchaser'; and 'holding' was defined as 'all the production units operated by the single producer and located within the geographical territory of the Community': Council Reg. (EEC) 3950/92, [1992] OJ L405/1, Art.9(c) and (d).

[126] Council Reg. (EEC) 1560/93, L154/30. For the current definition of 'producer' and 'holding' in the UK legislation, see the Dairy Produce Quotas Regs. 1994, S.I.1994 No.672, reg.2(1): in both cases the Community definition is adopted without qualification.

[127] Case 341/89, [1991] ECR I–25, at I–39. That a 'producer' could be a tenant had earlier been implicit in Case 5/88, *Wachauf* v. *Bundesamt für Ernährung und Forstwirtschaft*, [1989] ECR 2609, [1991] 1 CMLR 328. However, a landowner who has let out all his holding will not be a 'producer': Case 236/90, *Maier* v. *Freistaat Bayern*, [1992] ECR I–4483.

quantity in the name of the *partnership* rather than the name of the individual partners. This is consistent with the definition of 'producer', which includes 'a group of natural or legal persons'; and, moreover, the European Court has held that allocations should be made to the 'partnership as such'.[128] Further, in the case of *Herbrink* v. *Minister van Landbouw, Natuurbeheer en Visserij* the Advocate-General took care to distinguish between associations which enjoy their own legal personality and associations which do not, such as two farmers operating the holding together.[129] In the former circumstances the reference quantity was to be allocated to the association; and in the latter circumstances the two farmers would 'be jointly entitled to the reference quantity', but in their capacity as a group of persons — so satisfying one of the alternatives within the definition of 'producer'.[130] That having been said, he recognized the wide range of business structures which would fall to be considered by the national courts; and, in English law, particular difficulties have flowed from the fact that a partnership is not a distinct legal entity.[131] When faced by such questions, a practical approach was adopted in the case of *R.* v. *Dairy Produce Quota Tribunal for England and Wales*, ex parte *Atkinson*, the High Court believing it arguable that to qualify as a producer it was sufficient that an individual be a member of a partnership engaged in dairy farming.[132] However, the position becomes more complicated where, for example, the freehold or leasehold interest in the land is vested in only one of the partners, who makes the property available by licence to himself and the other partners for the purposes of the trade.[133] While in this event the partnership will generally be registered as the producer, under the current United Kingdom legislation there is a good argument that the grant of such a licence would be insufficient to effect a transfer of quota into its name.[134] That having been

---

[128] Case 84/90, *R.* v. *Ministry of Agriculture, Fisheries and Food*, ex p. *Dent*, [1992] ECR I–2009, [1992] 2 CMLR 597, at I–2036 (ECR) and 612 (CMLR) — this judgment being delivered in the context of allocations to farmers who had not received a reference quantity under the initial legislation through compliance with non-marketing or conversion undertakings. See also the Opinion of the Advocate-General in Case 86/90, *O'Brien* v. *Ireland*, [1992] ECR I–6251, [1993] 1 CMLR 489.

[129] Case 98/91, [1994] ECR I–223, [1994] 3 CMLR 645.

[130] Ibid., at I–241 (ECR) and 661–2 (CMLR).   [131] The Partnership Act 1890, s.1.

[132] (1985) 276 EG 1158. The point was not, however, fully explored.

[133] Such non-exclusive licences were found in, e.g., *Harrison-Broadley* v. *Smith*, [1964] 1 All ER 867, [1964] 1 WLR 456; and *Bahamas International Trust Co. Ltd.* v. *Threadgold*, [1974] 3 All ER 881, [1974] 1 WLR 1514. See also, e.g., Slatter, M. and Barr, W., *Farm Tenancies* (BSP Professional Books, Oxford, 1987), at 28–30; Muir Watt, J., *Agricultural Holdings* (13th edn., Sweet and Maxwell, London, 1987), at 11 and 17–18; Densham, H. A. C., *Scammell and Densham's Law of Agricultural Holdings* (7th edn., Butterworths, London, 1989), at 51–2 and 59–60; and Rodgers, C. P., *Agricultural Law* (Butterworths, London, 1991), at 41–2.

[134] The Dairy Produce Quotas Regs. 1994, S.I.1994 No.672, reg.7(6)(a)(i). Licences ceased to be a medium for quota transfer on the coming into force of the the Dairy Produce Quotas (Amendment) Regs. 1988, S.I.1988 No.534, reg.5 (1 Apr. 1988). For more detailed treatment of this aspects, see Ch.4.

said, for the landowning partner alone to be the registered producer would not accord naturally with the reality of the trading arrangements.[135]

It has also been common for a freehold owner or tenant to grant a 'licence' in favour of a company, the company then carrying on the trade and being registered as producer. However, should the company receive exclusive possession, in all probability the licence would be converted into a tenancy by section 2 of the Agricultural Holdings Act 1986 (the AHA 1986) — or under the common law principle enunciated in *Street* v. *Mountford*.[136] Should this occur, then the company would be constituted as tenant (or subtenant);[137] and, there having been a land transaction sufficient to effect a transfer of quota, registration in the name of the company would seem correct. However, where a subtenancy is inadvertently created in this way, the head-tenant may be open to action from his landlord for breach of any covenant against subletting.[138]

Similar considerations arise where the landowner makes available his property for the purposes of a share-farming agreement.[139] Again it is apprehended that such an agreement confers on the 'working farmer' no more than a licence to occupy jointly with the landowner;[140] and, accordingly, there is a strong argument that no land transaction has occurred sufficient to trigger a transfer of quota.[141] Likewise, there should be no transfer of quota on the termination of the agreement. In line with these

[135] Where a tenant is farming in partnership and the partnership is the registered producer, particular problems may be encountered when seeking compensation in respect of milk quota on the termination of the tenancy. For this aspect, see Ch.5, 4.2.2.

[136] See, e.g., Slatter, M. and Barr, W., *Farm Tenancies*, at 26–30; Muir Watt, J., *Agricultural Holdings*, at 9–22; Densham, H. A. C., *Scammell and Densham's Law of Agricultural Holdings*, at 30–2; and Rodgers, C. P., *Agricultural Law*, at 25–9 and 31–2. As highlighted by Densham, H. A. C., *Scammell and Densham's Law of Agricultural Holdings*, at 51, and Rodgers, C. P., *Agricultural Law*, at 31–2, it is not clear in what manner s.2 interacts with the common law as decided under *Street* v. *Mountford*, [1985] AC 809, [1985] 2 All ER 289, [1985] 2 WLR 877.

[137] See, e.g., *Snell* v. *Snell*, (1964) 191 EG 361; and, more recently, *Pennell* v. *Payne*, [1995] QB 192, [1995] 2 All ER 592, [1995] 2 WLR 261, [1995] 06 EG 152.

[138] See, e.g., *Snell* v. *Snell*, (1964) 191 EG 361; and, for a more benign view, *Pennell* v. *Payne*, [1995] QB 192, [1995] 2 All ER 592, [1995] 2 WLR 261, [1995] 06 EG 152.

[139] On share-farming, generally, see, e.g., Densham, H. A. C., *Scammell and Densham's Law of Agricultural Holdings*, at 53–4; Rodgers, C. P., *Agricultural Law*, at 39–41; and Stratton, R., Sydenham, A., and Baird, A., *Share Farming* (3rd edn., CLA Publications, London, 1992), *passim*. For a statutory definition of share-farming agreement, in the context of sheep annual premium quotas and suckler cow premium quotas, see the Sheep Annual Premium and Suckler Cow Premium Quotas (Amendment) Regs. 1993, S.I.1993 No.3036, reg.2(5): '"sharefarming agreement" means a farming contract made between the owner of land (including a person entitled for a term of years certain or other limited estate) and a farmer with no legal interest or charge in or over the land concerned which does not constitute the parties partners or employer and employee but merely contracting parties whose liablities remain separate, whose contributions are defined by the agreement between them, whose responsibility for planning and managing the farming enterprise is joint and whose rewards are an agreed share of the revenue of that enterprise; . . .'.

[140] See, e.g., *McCarthy* v. *Bence*, [1990] 17 EG 78 (Rodgers, C. P., [1991] Conv. 58–65).

[141] See, in particular, Stratton, R., Sydenham, A., and Baird, A., *Share Farming*, at 11–12.

arguments each party would be advised to have their own reference quantity and their own registration, the respective reference quantities being made available under the terms of the share-farming agreement for the duration of the agreement.[142] The position would seem somewhat clearer in the case of a genuine contracting agreement, under which, for example, the contractor provides services at a stipulated rate per hour. In these circumstances there can be little doubt that the landowner would be the 'producer' in whose name the reference quantity must be registered, the contractor enjoying no more than a right to go onto the land for the purpose of carrying out those services.

Some authority in this context is provided by the case of *Ballmann* v. *Hauptzollamt Osnabrück*.[143] The applicant owned a dairy farm and, as a producer in his own right, had been allocated a reference quantity corresponding to the milk production of approximately forty cows. Located on the farm there were twenty new stalls in a separate shed; and, since these were surplus to his requirements, he leased them to a tenant. At the time of the grant of the lease the tenant in question already enjoyed a reference quantity corresponding to the milk production of approximately twenty cows, established through of milk production on his own farm in 1983. Under the lease the applicant and the tenant provided separately for the feeding, milking, insemination, and veterinary treatment of their respective cows; and the milk was stored in separate tanks. However, the milking machinery and certain general facilities were shared. The regional authority gave notice that on these facts the tenant could no longer be regarded as a 'producer' and that his production should be set against that of the applicant landlord. By contrast, the European Court held that a tenant could indeed be a 'producer'; and that, as a rule, his production should be set against his own reference quantity rather than that of his landlord. However, this general rule would only apply where the tenant could demonstrate that he was operating on an independent basis the production units for whose operation he had taken out the lease. In particular, where certain facilities were shared with the landlord, the tenant would need to show that the milk he produced was stored and delivered separately. There can be no doubt that this judgment does highlight the possibility of two registered producers operating in close proximity.[144] That having been said, the relationship

---

[142] Moreover, if the landowner alone enjoys a reference quantity, the use of this by the working farmer could provide an indication that, in reality, the relationship is that of partnership rather than that of two separate business enterprises. If the working farmer does not himself enjoy the benefit of a reference quantity, this potential impasse could be cured by leasing: ibid.

[143] Case 341/89, [1991] ECR I–25.

[144] It is of note that the Court's concern was directed to ensuring the efficient administration of the milk quota system, the judgment emphasizing the need to ascertain with accuracy the amount of production to be set against each reference quantity.

between the parties would seem to have fallen short of a share-farming agreement: for example, there was no sharing of gross profits. In English law it might rather be interpreted as a lease of the twenty stalls, combined with a licence of the milking machinery and certain general facilities.

With regard to the second area of potential difficulty, no easy answer presents itself as to the extent to which a producer must be actively engaged in milk production. On the one hand, it has remained a central pillar of the milk quota system that reference quantities are allocated for the purposes of occupational activity.[145] Indeed, as a rule allocations were made to producers based upon *production* during the requisite reference year. Moreover, it has been seen that the definition of 'holding' refers to the production units 'operated' by the producer; and, in the context of the original transfer regulations, where one or several parts of a holding was sold, leased, or transferred by inheritance, the corresponding reference quantity was to be distributed among the producers 'operating' the holding.[146] On the other hand, there is evidence of increasing latitude towards farmers registered as producers who do not themselves carry out milk production (such farmers generally turning their reference quantities to account by leasing them for high prices to other producers).[147] Although these practices have been criticized,[148] infringement proceedings have been dropped; and, in particular, it is now open to producers to lease out all, rather than just part, of their reference quantities. In addition, as shall be

---

[145] See, e.g., the decisions of the European Court in Case 44/89, *Von Deetzen* v. *Hauptzollamt Oldenburg* ('*Von Deetzen II*'), [1991] ECR I–5119, [1994] 2 CMLR 487; and Case 2/92, *R.* v. *Ministry of Agriculture, Fisheries and Food*, ex p. *Bostock*, [1994] ECR I–955, [1994] 3 CMLR 547; and the decision of the German Administrative Court in *Re the Küchenhof Farm*, [1990] 2 CMLR 289. Moreover, the very word 'producer' would seem to indicate that the farmer should be engaged in milk production: see, e.g., the Opinion of the Advocate-General in Case 5/88, *Wachauf* v. *Bundesamt für Ernährung und Forstwirtschaft*, [1989] ECR 2609, [1991] 1 CMLR 328.

[146] Comm. Reg. (EEC) 1371/84, [1984] OJ L132/11, Art.5(2). Indeed, in the tax case of *Cottle* v. *Coldicott* the Special Commissioners interpreted this provision to the effect that, if part of a holding was sold to a person who was not a milk producer, then the reference quantity would not be transferred: [1995] SpC 40. Where an entire holding was sold, leased or transferred by inheritance, the corresponding reference quantity was to be transferred to the producer who 'takes over' the holding: ibid., Art.5(1). For the current Community legislation, see Council Reg. (EEC) 3950/92, [1992] OJ L405/1, Art.7(1): 'Reference quantities available on a holding shall be transferred with the holding in the case of sale, lease or transfer by inheritance to the producers taking it over . . .'. It may also be noted that the legislature felt it necessary to adopt an extended definition of 'producer' for the purposes of the subsequent allocations of reference quantities to farmers who had no representative production in their Member State's reference year owing to compliance with non-marketing or conversion undertakings: Council Reg. (EEC) 764/89, [1989] OJ L84/2. For a somewhat equivocal view as to whether such farmers could qualify as producers (without express legislative provision), see the Opinion of the Advocate-General in Case 120/86, *Mulder* v. *Minister van Landbouw en Visserij* ('*Mulder I*'), [1988] ECR 2321, [1989] 2 CMLR 1, at I–2239–40 (ECR) and 6 (CMLR).

[147] Such farmers are generally termed 'non-producing producers' or 'non-active quota holders'.

[148] See, e.g., Court of Auditors' *Special Report No.4/93 on the Implementation of the Quota System intended to control Milk Production*, [1994] OJ C12/1, at 4.39.

seen, the courts have interpreted the definition of the 'holding' operated by the producer in such a way that it may contain land not currently employed for milk production — for example, the arable part of a mixed holding. Perhaps the clearest illustration of the current attitude is provided by the confiscation provisions in Article 5 of Council Regulation (EEC) 3950/92.[149] These direct that the reference quantities of producers who have not marketed milk or milk products during a given milk year shall be forfeited and placed in the national reserve. That having been said, the effect of the sanction is materially weakened on two accounts. First, the producer may escape confiscation by merely entering into temporary transfers, there being no requirement that he use the reference quantity personally; and, secondly, the reference quantity may be restored within, in the case of the United Kingdom, six years from the beginning of the milk year of confiscation.[150] Accordingly, there are good reasons for effecting temporary transfers of quota which would otherwise remain unused during a given milk year; and, indeed, the general latitude extended to farmers not actively engaged in milk production may be seen as consistent with the fact that they provide a pool of quota to be exploited, for a price, by other farmers who may wish to commence milk production or expand existing production.[151]

Turning to the definition of 'holding', as a preliminary matter this must be clearly distinguished from the 'agricultural holding' as governed by the AHA 1986 and, indeed, from a 'farm business tenancy' as governed by the Agricultural Tenancies Act 1995 (the ATA 1995).[152] Two differences may be highlighted. First, a holding under the Community legislation may comprise freehold land, or tenanted land, or a mixture of both freehold land and tenanted land (the last mentioned being frequently termed a 'composite holding'). By contrast, an agricultural holding or a farm business tenancy is confined to tenanted land. Secondly, a holding under the Community legislation now in force extends to 'all production units operated by the single producer and located within the geographical territory of a Member

---

[149] [1992] OJ L405/1. For the UK implementing legislation, see now the Dairy Produce Quotas Regs. 1994, S.I.1994 No.672, reg.33, as amended by the Dairy Produce Quotas (Amendment) Regs. 1994, S.I.1994 No.2448.

[150] See also the Dairy Produce Quotas Regs. 1994, S.I.1994 No.672, reg.25(6) (concerning the preparation and maintenance of registers by the Intervention Board): 'In this regulation "direct seller" and "producer" include a person who occupies land with quota whether or not that person is engaged in the sale or delivery of dairy produce.'

[151] This role of non-active quota holders was highlighted in the *Third Report from the Agriculture Committee: Trading of Milk Quota* (Session 1994–5, H.C.512), at xiv. For a benign, purposive approach towards a mortgagee in possession who was making temporary transfers rather than selling or supplying milk or milk products, see *Harries* v. *Barclays Bank plc* (not yet reported, High Court, 20 Dec. 1995).

[152] For the definition of 'agricultural holding', see the AHA 1986, s.1. For the meaning of 'farm business tenancy', see the ATA 1995, s.1. The farm business tenancy régime has general application to tenancies granted on or after 1 September 1995, the grant of tenancies within the AHA 1986 being restricted on or after that date to the exceptional circumstances set out in s.4 of the ATA 1995.

State'.[153] By contrast, an agricultural holding or a farm business tenancy is confined to land comprised in a single contract of tenancy.[154] In the light of these major distinctions a holding under the Community legislation has frequently been termed a 'Euro-holding', to obviate risk of confusion.

As emphasized by the Advocate-General in *Wachauf* v. *Bundesamt für Ernährung und Forstwirtschaft*, the definition of 'holding' is a broad one.[155] More specifically, as confirmed by the European Court in the same case, the definition is wide enough to include tenanted land — even where the production units at the date of the grant have neither dairy cows nor the necessary facilities for milk production, and the agreement imposes no obligation to engage in milk production. The Court followed the Advocate-General, who rejected the view that, to qualify as a 'holding', the farm must be equipped and used directly and exclusively for milk production (or be destined for such use by, for example, a clause in the agreement to that effect). The decision was motivated in part by a concern to ensure that mixed farms fell within the scope of the legislation (and, in particular, the transfer rules). However, the Advocate-General did note that the inclusion of a reference to the 'producer' in the definition of 'holding' indicated that the 'farm must be engaged in milk production'; and, in the case of a mixed farm, he emphasized that milk production must be 'actually carried out on the holding'.[156]

Conflicting dicta have emerged from the United Kingdom national courts. In *Carson* v. *Cornwall County Council* the County Court, accepting the expert advice of Professor Usher to the arbitrator, held that land which the occupier was obliged by contract not to use for milk production could not be regarded as part of a holding.[157] By contrast, in *W.E. & R.A. Holdcroft* v. *Staffordshire County Council* the Court of Appeal took a more liberal approach.[158] The answers of the European Court in *Wachauf* were recited and it was stated that those answers rendered it plain that 'a farm is perfectly capable of being a holding even if the cows have all gone'.[159]

[153] Council Reg. (EEC) 3950/92, [1992] OJ L405/1, Art.9(d), as amended by Council Reg. (EEC) 1560/93 [1993] OJ L154/30. Until that amendment a 'holding' extended to all such production units within the Community.

[154] However, for a difficult case on this aspect (in the context of rent review), see *Trustees of J. W. Childers Deceased's Will Trust* v. *Anker*, [1995] EGCS 116.

[155] Case 5/88, [1989] ECR 2609, [1991] 1 CMLR 328.

[156] Ibid., at I–2626–7 (ECR) and 338–9 (CMLR).

[157] [1993] 03 EG 119. The case concerned the valuation of milk quota, the context being a claim for compensation on the termination of a tenancy. The tenant had, *inter alia*, adduced evidence based on prices paid for milk quota transferred through the medium of short-term occupation agreements, a key ingredient of such transactions being the contractual obligation on the transferee not to use the land for milk production (on which, see Ch.4, 3.).

[158] [1994] 28 EG 131.

[159] Ibid., at 134. However, in *Wachauf* the European Court was concerned to establish that tenanted land could qualify as a holding even though there had been no dairy use at the *commencement* of a lease.

Accordingly, the issue cannot be regarded as free from doubt. While there is a logic to the decision in *Carson* v. *Cornwall County Council*, it must be recognized that the broader interpretation remains the foundation of the many quota transfers still conducted through the medium of short-term tenancies.[160]

6. CHALLENGES TO THE IMPLEMENTING LEGISLATION

6.1. Challenges to the principle of production control

In the light of this complexity and, notably, the twin objectives of curbing milk production while at the same time continuing structural change, it was perhaps not surprising that the implementing legislation was subjected to frequent challenge in the courts. That having been said, the overarching principle of production control remained relatively immune from attack — consistent with earlier case law. For example, in *Hauer* v. *Rheinland-Pfalz* the European Court had accepted the validity of a temporary prohibition on the grant of permits to plant vines.[161] Although the right to property was enshrined in Community law in accordance with the constitution of Member States and the First Protocol to the European Convention for the Protection of Human Rights,[162] there could be restrictions on the use of property, provided that these restrictions corresponded to 'objectives of general interest pursued by the Community'. On the facts the restrictions satisfied this criterion; and, accordingly, with regard to the aim pursued they did not 'constitute a disproportionate and intolerable interference with the rights of the owner, impinging upon the very substance of the right to property'.[163] In particular, the Court emphasized the temporary nature of

[160] For interpretation of the expression 'holding' in the context of the allocations to farmers who failed to secure a reference quantity through compliance with non-marketing or conversion undertakings, see Case 86/90, *O'Brien* v. *Ireland*, [1992] ECR I–6251, [1993] 1 CMLR 489. It may be noted that the Advocate-General believed that the definition allowed 'the producer's "holding" at any given time to include the land then used by him for milk production': at I–6270 (ECR) and 499 (CMLR).

[161] Case 44/79, [1979] ECR 3727, [1980] 3 CMLR 42. See also Usher, J. A., 'Rights of Property: How fundamental?', (1980) 5 E.L.Rev. 209–12.

[162] See also Case 4/73, *Nold* v. *Commission*, [1974] ECR 491, [1974] 2 CMLR 338.

[163] Case 44/79, [1979] ECR 3727, [1980] 3 CMLR 42, at 3747 (ECR) and 66 (CMLR). See also, e.g., Case 5/88, *Wachauf* v. *Bundesamt für Ernährung und Forstwirtschaft*, [1989] ECR 2609, [1991] 1 CMLR 328; Case 177/90, *Kühn* v. *Landwirtshaftskammer Weser-Ems*, [1992] ECR I–35, [1992] 2 CMLR 242; and Case 280/93, *Germany* v. *Council*, [1994] ECR I–4973. It may be noted that the 'undeniable public interest' of the restriction was invoked in the implementing measure: Council Reg. (EEC) 1162/76, [1976] OJ L135/32, Preamble. Likewise it was subsequently stated that milk quotas should be imposed 'as a matter of overwhelming public interest': Council Reg. (EEC) 857/84, [1984] OJ L90/13, Preamble. On the general principles of Community law as applied in the agricultural context, see, e.g., Schmitthof, C. M., 'The Doctrines of Proportionality and Non-Discrimination', (1977) 2 E.L.Rev. 329–34; Vadja, C., 'Some Aspects of Judicial Review within the Common Agricultural Policy', (1979) 4

the prohibition, since the landowner retained the ability to dispose of the land or to put it to some other permitted use.[164] Further, throughout the Member States there was a social function to the right to property, as seen in legislation on agriculture, forestry, the water supply, the protection of the environment, and town and country planning — all of which imposed restrictions, sometimes appreciable, on the use of such right. Similar reasoning was employed to reject further arguments based on the freedom to pursue a profession. Of importance was the European Court's acceptance, as a matter of principle, that the Community has the power to restrict the exercise of the right to property in the context of the common organization of the market — and for the purposes of structural development. However, as seen, the measures enacted must not be disproportionate.

In the more directly analogous decision of *Stölting* v. *Hauptzollamt Hamburg-Jonas* a challenge to the validity of the co-responsibility levy was rejected, notwithstanding the adverse effect on the applicant, on the ground that it was justified as a matter of policy: the measure sought to stabilize the market, and, besides, its rate was not disproportionate.[165]

With regard to milk quotas themselves, an approach consistent with this earlier authority was adopted by Murphy J in the Irish case of *Lawlor* v. *Minister for Agriculture*.[166] In upholding the validity of the Community and national legislation, he affirmed that the superlevy regulations did not attempt to abolish the right of private property or the general right to transfer, bequeath, or inherit property. Rather, they delimited the exercise of certain rights of ownership, this delimitation being clearly aimed at reconciling the exercise of ownership rights with the exigencies of the common good. Accordingly, monetary compensation was not appropriate. Moreover, even if there was an infringement of the Constitution (and in the view of the judge there was not), the validity of the domestic legislation

---

E.L.Rev. 244–61, and 341–55; Neri, S., 'Le principe de proportionnalité dans la jurisprudence de la Cour relative au droit communautaire agricole', (1981) 4 R.T.D.E. 652–83; Wainwright, R., 'Legal Aspects of the Agricultural Policy in the European Community: General Principles of Law — Recent Developments', in Bates, St.J., Finnie, W., Usher, J. A., and Wildberg, H. (edd.), *In Memoriam J. D. B. Mitchell* (Sweet and Maxwell, London, 1983), at 163–70; Snyder, F. G., *Law of the Common Agricultural Policy*, at 38–9, and 56–60; Usher, J. A., *Legal Aspects of Agriculture in the European Community*, at 45–52; Sharpston, E., 'Legitimate Expectations and Economic Reality', (1990) 15 E.L.Rev. 103–60; Cardwell, M. N., 'General Principles of Community Law and Milk Quotas', (1992) 29 C.M.L.Rev. 723–47; and Barents, R., *The Agricultural Law of the EC*, at 315–61. See also, in the context of the Common Fisheries Policy, O'Reilly, J., 'Judicial Review and the Common Fisheries Policy in Community Law', in Curtin, D. and O'Keefe, D. (edd.), *Constitutional Adjudication in European Community and National Law: Essays for the Hon. Mr Justice T. F. O'Higgins* (Butterworths, Dublin, 1992), at 51–65.

[164] Cf., the written observations of the Council in Case 44/89, *Von Deetzen II*, [1991] ECR I–5119, at I–5129: in the Council's view, the issue was a 'straightforward limitation on the use of real property, namely the requirement of not making full use of a dairy holding's full reference production'.

[165] Case 138/78, [1979] ECR 713.

[166] [1990] 1 IR 356, [1988] ILRM 400, [1988] 3 CMLR 22.

would be saved on the basis that it was 'necessitated' by Community obligations.[167]

6.2. Challenges to detailed measures before the European Court

6.2.1. The variety of cases

The European Court has not only been prepared to accept restrictions on production in principle; it has also upheld many of the detailed measures within the framework of the milk quota system. Behind the decisions can be seen a determination to ensure the effectiveness of the superlevy and a sufficient degree of certainty in its implementation. Further, the European Court has, on several occasions, shown itself willing to preserve the discretion of the Community institutions when administering so complex an economic sector as the Common Agricultural Policy. None the less, the variety and ingenuity of the challenges has not left the detailed legislation unscathed. Certain cockpits of dispute may be highlighted: the provisions governing Formula A and Formula B; the provisions governing the reference years of Member States and producers; and the provisions governing development claims.

6.2.2. The provisions governing Formula A and Formula B

It was alleged in *Wuidart* v. *Laiterie coopérative eupenoise* that the choice granted to Member States between Formula A and Formula B gave rise to breach of the prohibition against discrimination among Community producers (as specifically enunciated in the agricultural context by Article 40(3) of the EC Treaty).[168] The Court held that, although production could be offset at the level of the purchaser under Formula B, it was reasonable for the Council to take the view that the advantage would be neutralized by the higher rate of levy applicable (i.e., 100 per cent of the target price for milk, rather than the 75 per cent rate applicable under Formula A).[169] In preserving the Community institutions' discretion, it was confirmed that

---

[167] See Arts.40 and 43 of the Irish Constitution.

[168] Joined Cases 267–285/88, [1990] ECR I–435. On Art.40(3) of the EC Treaty as a specific enunciation of the general principle prohibiting discrimination between Community producers, see, e.g., Joined Cases 117/76 and 16/77, *Albert Ruckdeschel and Co. and Hansa-Lagerhaus Ströh and Co.* v. *Hauptzollamt Hamburg-St. Annen* and *Diamalt AG* v. *Hauptzollamt Itzehoe*, [1977] ECR 1753, [1979] 2 CMLR 445; Joined Cases 124/76 and 20/77, *SA Moulins et Huileries de Pont-à-Mousson* v. *ONIC*, [1977] ECR 1795, [1979] 2 CMLR 445, and Joined Cases 201–202/85, *Klensch* v. *Secrétaire d'État à l'Agriculture et à la Viticulture*, [1986] ECR 3477, [1988] 1 CMLR 151. This position was confirmed in *Wuidart* v. *Laiterie coopérative eupenoise* itself, at I–480.

[169] However, all producers under Formula A were soon subject to superlevy at the full 100% rate: Council Reg. (EEC) 774/87, [1987] OJ L78/3. That having been said, the position was mitigated by the ability to reallocate the unutilized reference quantities of producers or

where the legislature 'is obliged, in connection with the adoption of rules, to assess their future effects, which cannot be accurately foreseen, its assessment is only open to criticism if it appears manifestly incorrect in the light of the information available to it at the adoption of the rules in question'.[170]

6.2.3. The provisions governing the reference years of Member States and producers

The reference year provisions were the subject of more frequent challenge before the European Court, much reliance again being placed on the general principles of Community law. These challenges related to implementation both at the level of the Member State and at the level of individual producers. In *Klensch* v. *Secrétaire d'État à l'Agriculture et à la Viticuture* the applicants contested *inter alia* Luxembourg's decision to retain 1981 as its reference year for the operation of Formula B (rather than follow the majority of Member States by opting for 1983).[171] It was alleged that such action was discriminatory, on the ground that it favoured the semi-State dairy Luxlait. In particular, deliveries to Luxlait were higher in 1981 than in 1982 or 1983, some producers since 1981 having ceased dairy farming or transferred to the applicants' dairies. The European Court emphasized that Article 40(3) of the EC Treaty applied not only to Community legislation, but also its implementation at the level of the Member State; and that Article 40(3) precluded any Member State adopting 1981 as its reference year if, owing to the specific conditions of that State's market and, in particular, the structure of the agricultural activities carried out on its territory, discrimination would arise between producers.

At the level of individual producers, the Court took a restrictive line regarding the exceptional events which might justify such producers having recourse to an alternative reference year from that adopted by their Member States. Indeed, this may be regarded as the forum for the most determined defence of the regulatory framework. In *Erpelding* v. *Secrétaire d'État à l'Agriculture et à la Viticulture* the applicant's herd had been severely affected from 1980 to 1984 by chronic mastitis and Canadian influenza.[172] The applicant therefore asked that account be taken of production in a year prior to 1981 or of a notional production calculated by extrapolation from production in earlier years. The claim failed, the Court refusing to

---

purchasers to producers or purchasers in the same region and, if necessary, in other regions: Council Reg. (EEC) 857/84, [1984] OJ L90/13, Art.4a, first inserted by Council Reg. (EEC) 590/85, [1985] OJ L68/1.

[170] [1990] ECR I–435, at 481. See also, e.g., Case 265/87, *Hermann Schräder HS Kraftfutter GmbH & Co. KG* v. *Hauptzollamt Gronau*, [1989] ECR 2237; and Case 280/93, *Germany* v. *Council*, [1994] ECR I–4973.

[171] Joined Cases 201–202/85, [1986] ECR 3477, [1988] 1 CMLR 151.

[172] Case 84/87, [1988] ECR 2647, [1989] 3 CMLR 493.

extend the range of available reference years beyond the 1981 to 1983 period. Most notably, it considered that 'the structure and the purpose of the rules concerned indicate that they contain an exhaustive list of the situations in which reference quantities or individual quantities may be granted and set out precise rules concerning the determination of those quantities',[173] a dictum which was to prove fatal to applicants on more than one occasion.

The Court further held that such an interpretation did not render the rules in question invalid. They were not incompatible with Article 39(1)(a) and (b) of the EC Treaty;[174] rather, by contributing to the stabilization of income, they had the very effect of securing a fair standard of living for the agricultural community. Nor did they breach the principles of proportionality or equality of treatment. With regard to the former, the Community institutions had indeed taken into account producers affected by an exceptional event in their Member State's reference year — by entitling these producers to opt for an alternative reference year. However, such choice could legitimately be restricted to another year within the 1981–3 period, in order to comply with 'the imperative requirements of legal certainty and the effectiveness of the additional levy system'.[175] With regard to the principle of equality of treatment, as specifically enunciated in Article 40(3) of the EC Treaty, again the range of alternative reference years had to be limited in the interests of both legal certainty and the effectiveness of the additional levy system. Accordingly, any difference in treatment was objectively justified.[176]

A similar line was adopted in *Leukhardt* v. *Hauptzollamt Reutlingen*, where the applicant sought to establish his individual reference quantity on the basis of production in 1980, his deliveries having decreased substantially after that date.[177] In addition to rejecting argument founded on

---

[173] Ibid., at 2671 (ECR) and 503 (CMLR).

[174] Under Art.39(1)(a) it is an objective of the Common Agricultural Policy 'to increase agricultural productivity by promoting technical progress and ensuring the rational development of agricultural production and the optimum utilisation of the factors of production, in particular labour'; and, as already indicated, under Art.39(1)(b) it is an objective 'thus to ensure a fair standard of living for the agricultural community, in particular by increasing the individual earnings of persons engaged in agriculture'. See also Case 280/93, *Germany* v. *Council*, [1994] ECR I–4973.

[175] Case 84/87, [1988] ECR 2647, [1989] 3 CMLR 493, at 2674 (ECR) and 505 (CMLR). The Court also noted that reallocations of unutilized reference quantities might be available to assist persons in the applicant's predicament.

[176] For objective justification of difference in treatment within the context of the Common Agricultural Policy, see, e.g., Joined Cases 201–202/85, *Klensch* v. *Secrétaire d'État à l'Agriculture et à la Viticulture*, [1986] ECR 3477, [1988] 1 CMLR 151; Joined Cases 267–285/88, *Wuidart* v. *Laiterie coopérative eupenoise*, [1990] ECR I–435; and Case 280/93, *Germany* v. *Council*, [1994] ECR I–4973.

[177] Case 113/88, [1989] ECR 1991, [1991] 1 CMLR 298. See also Case 67/89, *Berkenheide* v. *Hauptzollamt Münster*, [1990] ECR I–2615; and Case 85/90, *Dowling* v. *Ireland*, [1992] ECR I–5305, [1993] 1 CMLR 288.

Article 40(3) of the EC Treaty, the general principle of equality, and the right to enjoy property (as in *Erpelding* v. *Secrétaire d'État à l'Agriculture et à la Viticulture*), the Court also declared that there had been no infringement of the right to pursue a profession or occupation or of the general principles of legal certainty and protection of legitimate expectations. A critical factor, consistent with the European Court's earlier decisions, was a concern to preserve the Community legislature's wide discretion when it was required to evaluate a complex situation such as the Common Agricultural Policy.[178]

The European Court was equally inflexible where the applicant sought to extend not the range of reference years but the range of exceptional events justifying the choice of an alternative reference year by an individual producer. Thus, in *Kühn* v. *Landwirtschaftskammer Weser-Ems* the landlord of a German dairy farm failed to convince the Court that a change of tenant during the national reference year (1983) constituted such an exceptional event.[179] There was no doubt that the effect on production had been severe;[180] but, in line with its earlier decisions in *Erpelding* v. *Secrétaire d'État à l'Agriculture et à la Viticulture* and *Leukhardt* v. *Hauptzollamt Reutlingen*, the Court confirmed that the implementing regulations provided a comprehensive legislative framework for the allocation of reference quantities — and the applicant's specific situation was not addressed by that legislative framework. Moreover, such an interpretation was held not incompatible with the general principles of Community law, in particular the principle of legitimate expectations, the right of ownership, the freedom of producers to carry on a business, and the prohibition against discrimination laid down in Article 40(3) of the EC Treaty. Two aspects of the decision may be emphasized. First, in refusing to accept the argument based on the principle of legitimate expectations, the Court laid down that such principle can be pleaded against a Community provision 'only to the extent that the Community itself has previously created a situation which can give rise to a legitimate expectation'.[181] As shall be seen, this criterion proved critical to the success of the producers who failed to obtain a reference quantity as a result of compliance with Community non-market-

[178] See also Case 114/76, *Bela-Mühle Josef Bergmann KG* v. *Grows-Farm GmbH & Co. KG*, [1977] ECR 1211, [1979] 2 CMLR 83 (on skimmed milk powder); Joined Cases 279–280/84 and 285–286/84, *Firma Walter Rau Lebensmittelwerke* v. *Commission*, [1987] ECR 1069, [1988] 2 CMLR 704 (on the sale of intervention butter at reduced prices over the Christmas period); Case 331/88, *R.* v. *Minister of Agriculture, Fisheries and Food*, ex p. *FEDESA*, [1990] ECR I-4023, [1991] 1 CMLR 507 (on the prohibition of hormonal drugs); Joined Cases 267–285/88, *Wuidart* v. *Laiterie coopérative eupenoise*, [1990] ECR I-435; and Case 280/93, *Germany* v. *Council*, [1994] ECR I-4973.

[179] Case 177/90, [1992] ECR I-35, [1992] 2 CMLR 242. The applicant's claim failed notwithstanding that the Advocate-General's Opinion was in his favour.

[180] In 1981 deliveries amounted to 220,000 kg, in 1982 to 200,000 kg, but in 1983 to only 87,000 kg (of which 55,000 kg were produced by the departing tenant and 32,000 kg by the incoming tenant).

[181] Case 177/90, [1992] ECR I-35, [1992] 2 CMLR 242, at I-63 (ECR) and 259 (CMLR).

ing or conversion schemes.[182] Secondly, in rejecting argument based on the right to property and freedom to pursue an occupation, the Court followed closely the reasoning adopted earlier in the case of *Hauer* v. *Rheinland-Pfalz*.[183] Those rights were not absolute. Rather, they must be considered in relation to their social function; and, accordingly, restrictions may be imposed on their exercise, in particular in the context of a common organization of the markets, 'provided that those restrictions in fact correspond to objectives of general interest pursued by the Community and do not constitute, with regard to the aim pursued, a disproportionate and intolerable interference, impairing the very substance of those rights'.[184] The detailed rules in question, as part of the milk quota system designed to curb the surplus in the dairy sector, corresponded with objectives in the general interest. Moreover, they affected the substance of neither the right to property nor the freedom to pursue an occupation, since it was always open to the farmers in question to change to other forms of agricultural production. That many farmers did in fact adopt other forms of agricultural production may be judged by the growth of a surplus in the beef sector — such growth being actively encouraged by Community conversion schemes.[185]

Again, the applicant failed in *Le Nan* v. *Coopérative Laitière de Ploudaniel*.[186] He had purchased a farm in October 1983, during the course of France's reference year. However, he did not commence production (jointly with his father) until 1 April 1984. When the reference quantity was fixed at a low level on the basis of production by the previous occupier during the first half of the reference year, it was alleged *inter alia* that this constituted an exceptional event. Once again the Court held that a change in management did not suffice.

6.2.4. *The provisions governing development claims*

A further focus of litigation was the derogation in favour of certain producers who had adopted development plans.[187] The provisions as implemented by France were considered by the European Court in the case of *Cornée* v. *Coopérative agricole laitière de Loudéac and Laiterie coopérative du Trieux*.[188] More specifically, French producers contested the allocation under this head of a single fixed reference quantity, regardless of the objectives of the development plan. In line with its general approach, the Court

[182] See Case 120/86, *Mulder I*, [1988] ECR 2321, [1989] 2 CMLR 1; and Case 170/86, *Von Deetzen I*, [1988] ECR 2355, [1989] 2 CMLR 327. On this aspect, see further Ch.2, 4.

[183] Case 44/79, [1979] ECR 3727, [1980] 3 CMLR 42.

[184] Case 177/90, [1992] ECR I–35, [1992] 2 CMLR 242, at I–63–4 (ECR) and 258 (CMLR).

[185] See, e.g., Court of Auditors' *Special Report No.4/93 on the implementation of the Quota System intended to control Milk Production*, [1994] OJ C12/1, at 4.64.

[186] Case 189/92, [1994] ECR I–261, [1994] 3 CMLR 53.

[187] I.e., the derogation authorized under Council Reg. (EEC) 857/84, [1984] OJ L90/13, Art.3(1).

[188] Joined Cases 196–198/88, [1989] ECR 2309.

emphasized that, when implementing the common organization of an agricultural market, Member States must observe the prohibition against discrimination between producers laid down in Article 40(3) of the EC Treaty.
It also mentioned that the Member States had a discretion whether or not
to enact this derogation.[189] On the facts there was held to be a breach of
Article 40(3), not least because the Community regulation stipulated that,
where the development plan was still being implemented, any allocation of
a special reference quantity should take account of the production objectives. Accordingly, while the validity of the Community rules was confirmed, national implementation fell short of the strict compliance required.
That having been said, the applicants enjoyed less success in their further
claim that the national rules were invalid by reason that they excluded from
benefit producers whose milk deliveries exceeded 200,000 litres in 1983. In
the view of the Court, the Community legislation only imposed upon Member States an obligation to take into account production objectives: exact
proportionality was not necessary. Consequently, such a ceiling could be
justified on the basis of other criteria, notably social criteria.

Similar considerations arose in *Spronk* v. *Minister van Landbouw en
Visserij.*[190] In that case a Dutch farmer, who had already completed his
development plan, was allocated a special reference quantity less than his
actual production capacity. The Court decided that, if a Member State
chose to exercise its discretion to make available reference quantities to
such producers, then there was a requirement to take into consideration,
first, the level of production achieved in the year the development plan was
completed;[191] and, secondly, the principle of non-discrimination between
producers. On the facts the national implementing measures were found
broadly acceptable. For example, it was legitimate in the interests of administrative simplification to calculate the allocations on the basis of a fixed
quantity for each new stall constructed, so long as the quantity was objectively justified.[192]

6.3. Challenges to detailed measures in the United Kingdom

In the United Kingdom, as has been seen, adjudication upon the various
claims imposed a huge burden on the competent authorities. At the same
time the sheer volume of litigation threw up many issues of administrative

---

[189] As indicated, this particular aspect is now the subject of a challenge to the European
Court, whose decision is awaited: Case 63/93, *Duff* v. *Minister for Agriculture and Food*. See
also [1993] 2 CMLR 969, [1994] IJEL 247 (Irish High Court); [1993] 2 CMLR 969 (Irish
Supreme Court).
[190] Case 16/89, [1990] ECR I–3185, [1992] 1 CMLR 331.
[191] This factor was subject to the proviso that, where such production was not representative
of the new capacity, the allocation must bear some relation to the new capacity.
[192] One method of ensuring objective justification would be to fix the quantity by reference
to the average national production per stall.

law; and in this context may be highlighted the decision of the House of Lords in *Caswell* v. *Dairy Produce Quota Tribunal for England and Wales*, regarding delay in applying for judicial review.[193] Further, the Dairy Produce Quota Tribunal and courts were required to address detailed aspects of the national implementing regulations. For example, in *R.* v. *Dairy Produce Quota Tribunal for England and Wales*, ex parte *Wynn-Jones* the Court of Appeal considered *inter alia* whether the applicant had been sufficiently affected by the weather for him to establish his base year revision claim;[194] and in *R.* v. *Dairy Produce Quota Tribunal for England and Wales*, ex parte *Lifely* the High Court clarified, *inter alia*, the nature of the 'transaction' or 'arrangement' which might found an exceptional hardship claim, adopting a very broad interpretation.[195] With regulations operating at such a depth of complexity and detail, it was perhaps not surprising that negligence claims ensued against advisers.[196]

6.4. Conclusion

The overall picture which emerges from the decisions of both the European Court and the national courts is one of reluctance to shake the main structure of the milk quota system. Such an approach may be seen as consistent with a determination to buttress the Community's objective of curbing milk production 'as a matter of overwhelming public interest'.[197] Allied to this

---

[193] [1990] 2 All ER 434, [1990] 2 WLR 1320 (considering RSC Ord.53, r.4(1) and Supreme Court Act 1981, s.36(1)). On this matter, generally, see, e.g., Craig, P. P., *Administrative Law* (3rd edn., Sweet and Maxwell, London, 1994), at 585–6; Wade, H. W. R. and Forsyth, C. F., *Administrative Law* (7th edn., Clarendon Press, Oxford, 1994), at 678–9: and Hadfield, B., 'Delay in Applications for Judicial Review — a Hope Deferred?', [1990] CJQ 219–25. See also, e.g., *R.* v. *Dairy Produce Quota Tribunal for England and Wales*, ex p. *Davies*, [1987] 2 CMLR 399, (1987) 283 EG 463 (Dairy Produce Quota Tribunal must adequately explain its decision); and *R.* v. *Dairy Produce Quota Tribunal for England and Wales*, ex p. *S. Dimelow Farms*, *The Times*, 7 Nov. 1988 (power to permit amendment to claim). However, such issues fall outside the scope of this work.

[194] (1987) 283 EG 643. See, in particular, the Dairy Produce Quotas (Definition of Base Year Revision Claims) Regs. 1984, S.I.1984 No.1048.

[195] [1988] 27 EG 79. As noted above, the criteria for establishing such a claim in respect of wholesale quota included a requirement that the claimant prior to 2 Apr. 1984 had 'entered into, or become obliged to enter into, a transaction or made an arrangement . . . the reasonably expected outcome of which . . .' was 'a level of wholesale delivery of dairy produce in respect of which, or a substantial part of which, wholesale quota is not otherwise capable under these regulations of being allocated to him': the Dairy Produce Quotas Regs. 1984, S.I.1984 No.1047, Sch.2, para.17. The court held *inter alia* that a verbal agreement was sufficient. On the facts the evidence was consistent with the agreement having legal effect; and, even if it did not have such effect, it could still constitute the requisite 'transaction' or 'arrangement'. Another case in this context was *R.* v. *Dairy Produce Quota Tribunal for England and Wales*, ex p. *Atkinson*, (1985) 276 EG 1158, which again related to an exceptional hardship claim.

[196] See, e.g., *Stubbs* v. *Hunt & Wrigley*, [1992] 20 EG 107; and *Hood* v. *National Farmers Union*, [1992] 25 EG 135 (High Court); [1994] 01 EG 109 (Court of Appeal).

[197] Council Reg. (EEC) 857/84, [1984] OJ L90/13, Preamble. See also *Lawlor* v. *Minister for Agriculture*, [1990] 1 IR 356, [1988] ILRM 400, [1988] 3 CMLR 22; and Case 177/90, *Kühn* v. *Landwirtschaftskammer Weser-Ems*, [1992] ECR I–35, [1992] 2 CMLR 242.

may be detected a jealous preservation of the Community institutions freedom of action in administering a common market, with the institutions'
assessment of the effect of their measures only open to criticism when
manifestly incorrect. Further, many claims foundered through concern to
ensure both legal certainty and the effectiveness of the milk quota system.
In this regard, where the Community institutions had legislated so as to
provide flexibility, the courts were not lightly prepared to extend the circumstances in which reference quantities could be allocated: rather, it was
maintained, the structure and purpose of the rules indicated that they
contained an exhaustive list of the situations in which allocations could be
made.[198]

Accordingly, although enshrined in the Treaty and recognized in Council
Regulation (EEC) 856/84 itself, social objectives were frequently required
to give way — as, for example, the appeal to Articles 39(1)(a) and (b) failed
in *Erpelding* v. *Secrétaire d'État à l'Agriculture et à la Viticulture*.[199] This line
of reasoning was not confined to the original implementing legislation.
Thus, in *Spain* v. *Council*, Spain sought to resist a cut in its guaranteed total
quantity by relying on Article 39(1)(b).[200] In particular, the argument was
put forward that such a cut would have very adverse economic and social
consequences in certain regions where the milk sector accounted for approximately 70 per cent of agricultural activity, based upon small-scale
family farms. Rejecting this challenge, the European Court protected the
Community institutions' ability to allow any one or more of the Common
Agricultural Policy's objectives to take temporary priority; and in this context the Council attached greatest importance to the need to stabilize the
market.[201]

That having been said, complex and lengthy though the legislation might
have been, it totally failed to address the circumstances of one whole
category of farmers, namely those farmers who had no representative production in their Member State's reference year through compliance with
non-marketing or conversion schemes. The ensuing litigation, which materially affected the efficacy of the milk quota system as a whole, will be
considered in the course of the next chapter.

---

[198] See, e.g., Case 84/87, *Erpelding* v. *Secrétaire d'État à l'Agriculture et à la Viticulture*,
[1988] ECR 2647, [1989] 3 CMLR 493.
[199] Ibid.   [200] Case 203/86, [1988] ECR 4563.
[201] Art. 39(1)(c) of the EC Treaty. See also Case 5/73, *Balkan-Import-Export GmbH* v.
Hauptzollamt Berlin-Packhof, [1973] ECR 1091; Case 29/77, *Roquette Frères SA* v. *France*,
[1977] ECR 1835; Joined Cases 279–280 and 285–286, *Firma Walter Rau Lebensmittelwerke* v.
Commission, [1987] ECR 1069, [1988] 2 CMLR 704; Case 311/90, *Hierl* v. *Hauptzollamt
Regensberg*, [1992] ECR I–2061, [1992] 2 CMLR 445; and Case 280/93, *Germany* v. *Council*,
[1994] ECR I–4973.

2

The Continuing Surplus and Further Allocations

1. THE CONTINUING SURPLUS

Following the introduction of the milk quota system, production continued to outstrip demand. In large part this state of affairs may be attributed to reduced consumption: for example, during the period 1987 to 1990 it is estimated that consumption fell from 95 to 91 million tonnes, notwithstanding subsidized sales.[1] The level of deliveries to dairies also caused concern. After a significant fall in the 1984/5 milk year, this consistently exceeded the guaranteed total quantities of Member States.[2] To combat the surplus, a series of cuts were imposed. Council Regulation (EEC) 1335/86 provided that the guaranteed total quantities for wholesale quota be permanently reduced by 3 per cent over the 1987/8 and 1988/9 milk years.[3] As recited in the Preamble, the measure was justified by the need to strike a balance between supply and demand. Council Regulation (EEC) 1343/86 implemented a like reduction in respect of direct sales quota.[4]

In addition, a temporary withdrawal of wholesale quota was effected under Council Regulation (EEC) 775/87.[5] This amounted to 4 per cent in the 1987/8 milk year, rising to 5.5 per cent in the 1988/9 milk year.[6] When the dairy surplus proved resistant to even so large a reduction, the Council was again obliged to impose a permanent cut in wholesale quota, this time of 1 per cent as from the 1989/90 milk year.[7] However, the efficacy of the measure was blunted by fixing the percentage temporarily withdrawn at 4.5 per cent rather than 5.5 per cent, with the result that the level of reference quantities available to producers after the combination of permanent cuts

---

[1] See Court of Auditors, *Special Report No.4/93 on the Implementation of the Quota System intended to control Milk Production*, [1994] OJ C12/1, at 2.12–20 and Annexes VI and X–XII.

[2] See COM(86)645, at 7; and Court of Auditors, *Special Report No.4/93 on the Implementation of the Quota System intended to control Milk Production*, [1994] OJ C12/1, at Annexes IV–V.

[3] [1986] OJ L119/19.   [4] [1986] OJ L119/34.   [5] [1987] OJ L78/5.

[6] At the same time producers were granted compensation for the reference quantities temporarily withdrawn: Council Reg. (EEC) 775/87, [1987] OJ L78/5, Art.2. Moreover, the maximum level of compensation for those wishing to discontinue production definitively was raised from 4 ECU to 6 ECU per 100 kg: Council Reg. (EEC) 776/87, [1987] OJ L78/8. Cf., the solutions adopted in the cereal sector: see Usher, J. A., *Legal Aspects of Agriculture in the European Community*, at 75.

[7] Council Reg. (EEC) 3879/89, [1989] OJ L378/1.

and temporary withdrawals remained unaltered from the previous milk year.[8]

It was therefore perhaps not surprising that within a further two years the Council could once more recite that an endemic surplus pertained in the dairy sector — and that there was an 'imperative need' to achieve some degree of balance between supply and demand. Consequently, it required another permanent cut as from the 1991/2 milk year, amounting to 2 per cent of the guaranteed total quantity for both wholesale quota and direct sales quota.[9] By that date large scale reform of the milk quota system was approaching; and the next major adjustment to quota levels would constitute a part of that general reform.[10]

For the purposes of implementing such permanent cuts or temporary withdrawals, the legislation provided that the reference quantities be reduced proportionately, with no distinction being made between holdings operated on an industrial scale and holdings operated on a small, family-type basis. In the case of *Hierl* v. *Hauptzollamt Regensburg* the applicant disputed the validity of Council Regulation (EEC) 775/87, insofar as it provided for the imposition of the temporary withdrawals irrespective of the size of the individual producer's reference quantity.[11] It was alleged that this infringed Article 39 of the EC Treaty and the principle of equal treatment. More specifically, the applicant had received, an initial reference quantity of only 17,000kg; and he claimed that, farming on so limited a scale, he was less able to compensate for the withdrawal of part of his reference quantity by either reducing his purchases of feeding-stuffs or intensifying other forms of production. In finding that there was no breach of Article 39, the Court emphasized that the temporary withdrawal was designed to secure a reasonable balance between supply and demand; and thus, by contributing to the stabilization of the markets, it fell squarely within Article 39(1)(c). The fact that the temporary withdrawal was accompanied by the payment of compensation also confirmed that the rules were not disproportionate. Moreover, even though there might be a temporary lowering of the standard of living of the applicant and his family (notwithstanding the compensation), such an effect had to be accepted. As seen earlier, the Community institutions could accord one of the objectives of the Common Agricultural Policy temporary priority;[12] and the legislature enjoyed a broad discretion corresponding to the political responsibilities imposed upon it.[13] Likewise, on the evidence, there was no apparent breach

[8] Council Reg. (EEC) 3882/89, [1989] OJ L378/6. Both these measures formed part of the 'Nallet Package', considered at Ch.2, 3 (Nallet being the French Agriculture Minister).
[9] Council Reg. (EEC) 1630/91, [1991] OJ L150/19, (wholesale quota); and Council Reg. (EEC) 1635/91, [1991] OJ L150/28, (direct sales quota).
[10] See Ch.3.   [11] Case 311/90, [1992] ECR I–2061, [1992] 2 CMLR 445.
[12] See, e.g., Case 203/86, *Spain* v. *Council*, [1988] ECR 4563.
[13] See, e.g., Joined Cases 267–285/88, *Wuidart* v. *Laiterie coopérative eupenoise*, [1990] I–435.

of the principle of equal treatment (enshrined in Article 40(3) of the EC Treaty); and, in any event, the rules were based on objective criteria formulated to meet the needs of the general common organization of the market. Accordingly, while in the past the Commission had countenanced a special levy on milk from intensive farms,[14] once the decision had been taken to employ milk quotas as the primary means of curbing production, the European Court was prepared to uphold reductions imposed without distinction upon all producers.

2. FURTHER ALLOCATIONS: GENERAL

The decline in dairy consumption and the high level of deliveries without doubt proved key factors in this endemic surplus. However, as from the 1989/90 milk year, the position was considerably exacerbated by substantial increases in the sum of reference quantities within the Community as a whole, for the purposes of meeting two further sets of allocations to producers. The first set was made under the umbrella of the 'Nallet Package' (named after the French Agriculture Minister). In addition to adjusting the level of permanent cuts and temporary withdrawals, this series of measures provided that Member States could allocate further reference quantities taking into account 'the special situation of certain producers'.[15] The second set of allocations was that, already mentioned, in favour of farmers who had no representative production during their Member State's reference year as a result of compliance with Community non-marketing or conversion schemes. Reference quantities allocated under this head have frequently been referred to as 'SLOM quota'.[16] The legal aspects of SLOM quota mark a major contribution to Community jurisprudence, the consequences extending far outside the confines of the milk quota system. Most notably, this has been the context for important judgments on both the general principle of legitimate expectations and the non-contractual liability of the Community institutions.

[14] See, e.g., COM(83)500, at 19–20.

[15] Council Reg. (EEC) 3880/89, [1989] OJ L378/3, Preamble.

[16] The Community legislation throughout speaks of 'special reference quantities'; and this is echoed in the UK statutory instruments, which employ the term 'special quota'. However, there has been general adoption of the expression 'SLOM quota', which has the advantage of distinguishing such allocations from others made in special circumstances: in particular, SLOM quota has been subject to restrictions on transfer which do not apply to any other allocations. There are various views as to the exact origins of the Dutch acronym 'SLOM'. For three possibilities, see Court of Auditors, *Special Report No.4/93 on the Implementation of the Quota System intended to control Milk Production*, [1994] OJ C12/1, at 4.10; Gehrke, H., *The Implementation of the EC Milk Quota Regulations in British, French and German Law*, at 39; and Brown, L. N. and Kennedy, T., *The Court of Justice of the European Communities*, at 93. Moreover, the expression has been freely used before the European Court: see, e.g., the Opinion of the Advocate-General in Case 98/91, *Herbrink* v. *Minister van Landbouw, Natuurbeheer en Visserij*, [1994] ECR I–223, [1994] 3 CMLR 645.

However, before considering these in greater detail, it may also be mentioned that in 1991 the need also arose to find reference quantities for producers making direct sales of such items as yoghurt and ice-cream. The catalyst was clarification of Council Regulation (EEC) 857/84.[17] Until that point it had been understood that the milk quota system extended only to milk and other milk products defined as 'cream, butter and cheese'.[18] However, Council Regulation (EEC) 306/91 specified that this definition of other milk products was non-restrictive. Accordingly, it embraced such items as yoghurt and ice-cream.[19] The United Kingdom succeeded in finding the necessary reference quantities from the national reserve;[20] and allocations of both 'primary quota' and 'secondary quota' were made under the Dairy Produce Quotas Regulations 1991.[21] The former category was available for established producers, the latter for producers who could show that before 1 March 1991 they were committed to developing production of such 'additional milk products'.[22]

3. FURTHER ALLOCATIONS: THE NALLET PACKAGE

As mentioned, under the Nallet Package the wholesale quota guaranteed total quantities of Member States were cut on a permanent basis by 1 per cent. However, as also mentioned, the effect was cancelled by a parallel reduction in the amount of quota temporarily withdrawn; and, furthermore, the 1 per cent of quota so cut was added to the Community reserve, available for distribution by Member States taking account of the special situations of certain producers.[23] There was considerable evidence of a social element to this legislation, with the categories of producers who might benefit specifically including new entrants and producers whose ref-

---

[17] [1984] OJ L90/13.   [18] Ibid., Art.12(b).   [19] [1991] OJ L37/4.

[20] In order to augment the national reserve, the reference quantities of individual producers had been cut by a greater amount than the 2% required under Council Reg. (EEC) 1630/91, [1991] OJ L150/19. As a result the national reserve was sufficient to meet these extra demands: see, e.g., Federation of UK Milk Marketing Boards, *United Kingdom Dairy Facts and Figures: 1992 Edition* (Thames Ditton, Surrey, 1992), at 63.

[21] S.I.1991 No.2232, reg.24 and Sch.9.

[22] The latter category, accordingly, may be compared with development claims and exceptional hardship claims under the original legislation. Further echoes of the original legislation may be detected in the mechanism for setting the base quantity of primary quota. This was fixed at 90% of the quantity of additional milk products sold during either 1988, 1989, or 1990, whichever quantity was (in the opinion of the Minister) the greatest. For a disputed claim for secondary quota, see *R.* v. *Dairy Produce Quota Tribunal for England and Wales*, ex p. *P. A. Cooper and Sons*, [1993] 19 EG 138.

[23] Council Reg. (EEC) 3880/89, [1989] OJ L378/3, inserting a new Art.3b into Council Reg. (EEC) 857/84, [1984] OJ L90/13; and Council Reg. (EEC) 3881/89, [1989] OJ L178/5.

erence quantities were less than or equal to 60,000 kg.[24] At the same time the measures may be seen as promoting beneficial structural change.

In the event the Nallet Package caused the Community reserve to be increased by some 1,039,885.740 tonnes for the 1989/90 milk year.[25] Of this total the United Kingdom received 153,295.740 tonnes.[26] In England and Wales the Community regulations were implemented by applying the reference quantities available for the benefit of producers who had not yet been allocated their development awards in full; and for the benefit of family-type holdings where milk production constituted a significant activity. In the latter case individual reference quantities could be brought up to 200,000 litres.[27] Again no advantage was taken of the opportunity to assist new entrants. The Ministry of Agriculture, Fisheries, and Food did in fact submit such measures for approval; but the reference quantities originally destined for new entrants passed to producers on family-type holdings. In *R. v. Ministry of Agriculture, Fisheries and Food*, ex parte *Lomax* this action was held valid.[28] The judge felt that the Commission was not concerned which categories of producer received allocations, so long as those which did were approved.

4. FURTHER ALLOCATIONS: SLOM QUOTA

4.1. *Mulder I* and *Von Deetzen I*

As indicated, the introduction of the co-responsibility levy was not the only measure taken in 1977 to curb the surplus in the dairy sector. Alongside that levy were instituted non-marketing and conversion schemes 'to encourage the trend noted among certain groups of holdings in the Community to cease milk production, or the marketing of milk and milk

---

[24] Council Reg. (EEC) 3880/89, [1989] OJ L378/3. See also Council Reg. (EEC) 1183/90, [1990] OJ L119/17 (authorizing Member States to allocate to small producers reference quantities freed through cessation schemes).

[25] Council Reg. (EEC) 3881/89, [1989] OJ L378/5. See also Council Reg. (EEC) 1184/90, [1990] OJ L119/30 (for the 1990/1 milk year); and Council Reg. (EEC) 1636/91, [1991] OJ L150/29 (for the 1991/2 milk year).

[26] Comm. Reg. (EEC) 652/90, [1992] OJ L71/14 (for the 1989/90 milk year); Comm. Reg. (EEC) 2138/90, [1990] OJ L195/23 (for the 1990/1 milk year); and Comm. Reg. (EEC) 2061/91, [1991] OJ L187/35 (for the 1991/2 milk year).

[27] Such a scheme of allocations was approved by Comm. Dec. of 12 March 1990, [1990] OJ L76/26. For implementation in the UK, see the Dairy Produce Quotas Regs. 1989, S.I.1989 No.380, as amended by the Dairy Produce Quotas (Amendment) (No.2) Regs. 1990, S.I.1990 No.664, the Dairy Produce Quotas (Amendment) (No.3) Regs. 1990, S.I.1990 No.784, and the Dairy Produce Quotas (Amendment) Regs. 1991, S.I.1991 No.832. In Scotland and Northern Ireland there was made available a small producer supplementary development provision; in Northern Ireland an exceptional hardship provision; and in remote areas of Scotland and the Isles of Scilly a remote areas provision.

[28] [1992] 12 EG 146.

products'.[29] Under the non-marketing scheme, producers undertook, *inter alia*, not to market milk or milk products from their holdings for a period of five years. Under the conversion scheme, producers undertook, *inter alia*, to convert their holdings to meat production for a period of four years. A very considerable number of farmers took advantage of these provisions — approximately 123,000 in total throughout the Community.[30] Although the schemes were closed to new applications in 1980, there were still many producers who were complying with their undertakings during the reference year of their Member State.[31] Accordingly, they had no representative production and received no right to an allocation.

The validity of the implementing legislation was challenged in *Mulder I* and *Von Deetzen I*.[32] As a preliminary matter, the European Court was required to decide the extent to which the implementing legislation addressed the predicament of the applicants. It has already been seen that the European Court has, as a rule, shown itself astute to preserve the exhaustive nature of the legislation governing allocations of reference quantities. Within these confines the Court felt that the provisions in question did not necessarily assist producers such as the applicants: their particular circumstances were not expressly addressed and, besides, there was no guarantee that they would fall within one or more of the situations in which special, specific, or additional reference quantities could be allocated. By way of example the Court in *Mulder I* cited producers whose non-marketing undertakings expired after their Member State's reference year and who had not entered into a development plan. Moreover, Member States could only take account of the situations envisaged in Articles 3 and 4 of Council Regulation (EEC) 857/84 within the limits of the quantities available for that purpose.[33] As a result there was no certainty that all producers who had entered into non-marketing or conversion schemes would receive a reference quantity.

On this basis it was claimed in *Mulder I* that Council Regulation (EEC) 857/84 was invalid for breach of the general principles of Community law. Particular recourse was had to the principle of legal certainty, the principle of proportionality, the right to enjoy property, the prohibition of discrimination between producers laid down in Article 40(3) of the Treaty, and the

---

[29] Council Reg. (EEC) 1078/77, [1977] OJ L131/1, Preamble. See also Comm. Reg. (EEC) 1307/77, [1977] OJ L150/24.

[30] Written Answer given by Mr MacSharry on behalf of the Commission to a question from Mr Hume: [1990] OJ C93/26.

[31] The schemes were closed to new applicants by Council Reg. (EEC) 1365/80, [1980] OJ L140/18.

[32] Case 120/86, [1988] ECR 2321, [1989] 2 CMLR 1; and Case 170/86, [1988] ECR 2355, [1989] 2 CMLR 327. For a discussion of the SLOM litigation and allocations, see *Milk Quotas: M2/93* (Country Landowners Association, London, 1993), at 23–6.

[33] [1984] OJ L90/13, Art.5.

prohibition of misuse of powers. The grounds for invalidity in *Von Deetzen I* were similarly based on the general principles of Community law, with the greatest emphasis on the protection of legitimate expectations. In finding for the applicants in both cases, the Court stated that the legislation was partially invalid for breach of the principle of protection of legitimate expectations. Critical to the decision was the fact that, where a producer 'has been encouraged by a Community measure to suspend marketing for a limited period in the general interest and against payment of a premium he may legitimately expect not to be subject, upon the expiry of his undertaking, to restrictions which specifically affect him precisely because he availed himself of the possibilities offered by the Community provisions'. Since the Court had already adjudged that producers in the circumstances of the applicants could not be sure of allocations of reference quantities under the existing legislation, they were indeed subject to such restrictions and Council Regulation (EEC) 857/84 was to that extent invalid. Further, in both cases the Court pointed out that there was nothing in Council Regulation (EEC) 1078/77 to show that the non-marketing undertaking might lead to permanent exclusion from the dairy sector. However, it was accepted that a producer who had entered into such schemes could not 'legitimately expect to be able to resume production under the same conditions as those which previously applied and not to be subject to any rules of the market or structural policy adopted in the mean time'.[34] In *Mulder I* the Advocate-General believed that there would also be a breach of the prohibition against discrimination. For example, such producers were in a different position to those who had never produ ced milk or who had given up production for reasons unconnected with Community schemes.[35] The

---

[34] Case 120/86, [1988] ECR 2321, [1989] 2 CMLR 1, at 2352 (ECR) and 15 (CMLR); and Case 170/86, [1988] ECR 2355, [1989] 2 CMLR 327, at 2372 (ECR) and 333 (CMLR). This approach was subsequently followed in Joined Cases 196–198/88, *Cornée* v. *Coopérative agricole laitière de Loudéac and Laiterie coopérative du Trieux*, [1989] ECR 2309. Among the matters for consideration by the European Court was the validity of French reductions in reference quantities in the 1985/6 milk year. It was held that, if such reductions were targeted at producers who had adopted development plans, then there would be a breach of the principle of protection of legitimate expectations. However, producers who had adopted development plans could not rely on the principle of protection of legitimate expectations, provided that the reductions were permitted by the Community rules and did not affect them specifically as a category of producer. The Court also held that the implementation of a development plan approved by the national authorities did not confer a right to produce milk corresponding to the plan's objectives free from subsequent Community rules, such as the milk quota system. See also Case 84/78, *Augelo Tomadini Snc* v. *Amministrazione delle Finanze dello Stato*, [1979] ECR 1801, [1980] 2 CMLR 573.

[35] That having been said, in the absence of discrimination or the breach of the principle of protection of legitimate expectations, he was not persuaded that there was a misuse of power or violation of the right to property; and arguments based on non-discrimination were preferred to those based on proportionality. Likewise, in *Von Deetzen I* he did not accept that the applicant's liberty to exercise a profession had been affected, for the reasons given in Case 44/79, *Hauer* v. *Rheinland-Pfalz*, [1979] ECR 3727, [1980] 3 CMLR 42.

Court, none the less, restricted the reasoning behind its decision to breach of the principle of protection of legitimate expectations, not finding it necessary to consider other arguments.

These judgments stand in stark contrast to the European Court's general approach to the implementing legislation — in *Erpelding* v. *Secrétaire d'État à l'Agriculture et à la Viticulture*,[36] *Leukhardt* v. *Hauptzollamt Reutlingen*,[37] and *Kühn* v. *Landwirtschaftskammer Weser–Ems*[38] that implementing legislation, with its inherent flexibility, being found to provide an exhaustive list of the situations in which reference quantities could be allocated. However, in *Mulder I* and *Von Deetzen I* there would seem to have been two principal differentiating features. First, the SLOM producers had been actively encouraged by the Community legislation to participate in the non-marketing and conversion schemes, rendering it the more inequitable that they should suffer precisely because of that participation. As the Advocate-General stated in *Mulder I*, they had so acted in the common interest; and, further, a decision to the contrary might have created a reluctance to participate in future non-marketing and conversion schemes. Secondly, the schemes had always been held out to be temporary; and yet their effect under the implementing regulations was total exclusion from the milk quota system.[39] As again stated by the Advocate-General in *Mulder I*, this crossed the line 'between what is merely "hard business luck" and what is unreasonable'.[40]

4.2. SLOM 1 allocations

Following the judgments in *Mulder I* and *Von Deetzen I* the Community institutions provided for the first round of allocations to SLOM producers, which came to be known as 'SLOM 1 quota'.[41] In the United Kingdom there were some 590 applicants, who received 128 million litres of quota.[42] The

[36] Case 84/87, [1988] ECR 2647, [1989] 3 CMLR 493.

[37] Case 113/88, [1989] ECR 1991, [1991] 1 CMLR 298.

[38] Case 177/90, [1992] ECR I-35, [1992] 2 CMLR 242.

[39] Cf., Case 44/79, *Hauer* v. *Rheinland-Pfalz*, [1979] ECR 3727, [1980] 3 CMLR 42 (where the Court upheld a temporary prohibition on the grant of new permits to plant vines).

[40] Case 120/86, [1988] ECR 2321, [1989] 2 CMLR 1, at 2341 (ECR) and 8 (CMLR). On these points, generally, see, e.g., Sharpston, E., 'Legitimate Expectations and Economic Reality', (1990) 15 E.L.Rev. 103–60; and Craig, P. P., 'Legitimate Expectations: a Conceptual Analysis', (1992) 108 L.Q.R. 79–98. For a clear statement of the operation of the principle of protection of legitimate expectations in this context, see the Opinion of the Advocate-General in Case 264/90, *Wehrs* v. *Hauptzollamt Lüneberg*, [1992] ECR I–6285.

[41] As already mentioned, the Community and UK legislation refer respectively to 'special reference quantities' and 'special quota' rather than 'SLOM quota'; but 'SLOM 1 quota' has the particular merit that it permits easy distinction between allocations made pursuant to *Mulder I* and *Von Deetzen I* and allocations made pursuant to later judgments ('SLOM 2 quota' and 'SLOM 3 quota').

[42] Milk Marketing Board, *Five Years of Milk Quotas: a Progress Report*, at 29.

legislation passed to remedy the partial invalidity exposed by *Mulder I* and *Von Deetzen I* was Council Regulation (EEC) 764/89,[43] which inserted a new Article 3a into Council Regulation (EEC) 857/84.[44] However, at the same time as making SLOM 1 quota available, the detailed provisions also imposed severe restrictions. Five aspects may be highlighted: the link with production; the '60 per cent rule'; the cut-off date; the provision against double counting; and the restrictions on transfer.

Turning first to the link with production, this was imposed with a view to limiting allocations to farmers who had both a *bona fide* intent and the capacity to return to milk production. For example, eligibility was denied to farmers who had ceased farming or transferred the whole of their dairy enterprise prior to the end of the non-marketing or conversion period.[45] Moreover, allocations were made on a provisional basis, only becoming definitive once a farmer had proved that he had actually resumed direct sales and/or deliveries within two years of 29 March 1989; and that in the previous twelve months such direct sales and/or deliveries had reached at least 80 per cent of the provisional allocation. Failure to achieve that target resulted in the entire provisional allocation being returned to the Community reserve. Where a producer met the stipulated conditions, he was retroactively exempted from superlevy on production for the period to 31 March 1989. However, this 'amnesty' was not available where he had resumed production but failed to satisfy the conditions, a rule held valid in *Herbrink* v. *Minister van Landbouw, Natuurbeheer en Visserij*.[46] Commission Regulation (EEC) 1033/89 required the farmers to prove in addition that they continued to operate, in whole or in part, the same holdings as those they operated at the time that their premium applications were approved.[47]

Secondly, entitlement was restricted to 60 per cent of the amount of milk sold or delivered during the twelve months preceding the month in which the application for the non-marketing or conversion scheme was made. This restriction has frequently been referred to as the '60 per cent rule'.

Thirdly, a cut-off date was imposed, only those producers whose non-marketing or conversion schemes expired after 31 December 1983 being able to apply.[48] The Council would subsequently seek to justify this provision before the European Court on the ground that farmers whose under-

---

[43] [1989] OJ L84/2. See also Comm. Reg. (EEC) 1033/89, [1989] OJ L110/27. Cf., Art.3b, inserted into Council Reg. (EEC) 857/84 to implement the allocations under the Nallet Package.

[44] [1984] OJ L90/13.

[45] 'Farming' for these purposes was defined in Council Reg. (EEC) 1078/77, [1977] OJ L131/1, Art.2(3) and (4).

[46] Case 98/91, [1994] ECR I–223, [1994] 3 CMLR 645.   [47] [1989] OJ L110/27.

[48] Exceptionally, the cut-off date was fixed at 30 Sept. 1983 in Member States where the milk collection in the months Apr. to Sept. was at least twice that of the months Oct. to Mar. in the following year. This exception did not apply in the UK.

takings expired before that date could lawfully have resumed production but exercised a choice not to do so.[49]

Fourthly, where a producer had already obtained a reference quantity under the provisions which addressed 'special situations' in Article 3(1) and (2) of Council Regulation (EEC) 857/84 and/or under the restructuring provisions in Article 4(1)(b) and (c) of the same regulation, then the allocation of SLOM 1 quota fell to be reduced by that amount.

Fifthly, SLOM 1 quota could not be the subject of a temporary transfer; and, if the holding was sold or leased prior to 31 March 1992, the SLOM 1 quota would be returned to the Community reserve (proportionately on the basis of the feed-crop area if only part of the holding was sold or leased). However, different rules applied where the holding was transferred by inheritance or by any similar transaction. In such circumstances the SLOM 1 quota passed with the holding, provided that the transferee undertook in writing to comply with the undertakings of his predecessor.[50]

The rationale behind these restrictions was clearly stated in the Preamble to Council Regulation (EEC) 764/89.[51] Quota by that date having already acquired a substantial value, it was not the purpose of the legislation to confer an 'undue advantage' on the producers concerned. Accordingly, it was appropriate to oblige the applicants to show that they had both the intention and the capacity to resume milk production; and to limit the transfer of the reference quantities so allocated. In this the regulation may be seen as consistent with the developing hostility on the part of the legislature towards producers whose primary motivation was the realization of the value of quota rather than milk production itself.

Further, the Community institutions were concerned to limit the amount of quota required to meet the claims of the SLOM producers in the light of 'the overriding necessity of not jeopardizing the fragile stability that currently obtains' in the market;[52] and great weight was attached to this ground for the purposes of justifying the 60 per cent rule. Moreover, in the subsequent cases of *Spagl* v. *Hauptzollamt Rosenheim* and *Pastätter* v. *Hauptzollamt Bad Reichenhall* the Commission submitted an estimate that without the 60 per cent rule some 1 million tonnes of SLOM 1 quota would have been required to meet the various claims.[53] None the less, even with the 60 per cent rule and the other restrictive measures, it proved necessary to increase the Community reserve by 600,000 tonnes, according to the Community legislature the largest quantity compatible with the objective of

---

[49] Case 189/89, *Spagl* v. *Hauptzollamt Rosenheim*, [1990] ECR I–4539.

[50] Comm. Reg. (EEC) 1033/89, [1989] OJ L110/27. For first implementation of this provision in the UK, see the Dairy Produce Quotas Regs. 1989, S.I.1989 No.380, as amended by the Dairy Produce Quotas (Amendment) Regs. 1990, S.I.1990 No.132.

[51] [1989] OJ L84/2.   [52] Ibid.

[53] Case 189/89, [1990] ECR I–4539; and Case 217/89, [1990] ECR I–4585.

the scheme short of cutting the reference quantities of farmers who had remained in production throughout.[54] When taken together with the increase under the Nallet Package, the overall effect on the level of reference quantities was very significant. Indeed, the Community reserve rose from 443,000 tonnes to 2,082,885.740 tonnes for the 1989/90 milk year;[55] and the scale of the problems presented may also be judged by the sheer number of applicants, some 13,137 in total.[56]

On the distribution of this 600,000 tonnes, the United Kingdom received some 160,289.521 tonnes, a greater amount than any other Member State.[57]

4.3. SLOM 2 and SLOM 3 allocations

4.3.1. General

The amendments effected by Council Regulation (EEC) 764/89 and Commission Regulation (EEC) 1033/89 were themselves challenged and held to be partially invalid. This led to two further rounds of allocations, respectively SLOM 2 quota and SLOM 3 quota.

4.3.2. SLOM 2 allocations

SLOM 2 quota became available consequent upon three decisions of the European Court, *Spagl* v. *Hauptzollamt Rosenheim, Pastätter* v. *Hauptzollamt Bad Reichenhall*, and *Rauh* v. *Hauptzollamt Nürnberg-Fürth*.[58] *Spagl* and *Pastätter* may conveniently be considered together, since they both addressed the validity of the 60 per cent rule. *Spagl* also addressed the validity of the cut-off date. The Court held that the 60 per cent rule breached the principle of protection of legitimate expectations. A reduction of 40 per cent in SLOM 1 allocations was more than twice the highest total of permanent cuts and temporary withdrawals imposed on producers who had continued production and received reference quantities under the normal rules. Accordingly, the restriction specifically affected the

---

[54] Ibid. The 600,000 tonnes for SLOM 1 producers was added to the Community reserve by Council Reg. (EEC) 766/89, [1989] OJ L84/6. See also, Lawrence, G., 'Milk Superlevy: the Community System', in *Milk Quotas: Law and Practice: Papers from the ICEL Conference — June 1989*, 1–10, at 5.

[55] See Council Reg. (EEC) 776/84, [1989] OJ L84/6; and Council Reg. (EEC) 3881/89, [1989] OJ L378/5.

[56] Written Answer given by Mr MacSharry on behalf of the Commission to a question from Mr Hume: [1990] OJ C93/26.

[57] Comm. Reg. (EEC) 3835/89, [1989] OJ L372/27. This figure later rose to 191,215 tonnes for the 1991/2 milk year 35: Comm. Reg. (EEC) 2061/91, [1991] OJ L187/35.

[58] Case 189/89, [1990] ECR I–4539; Case 217/89, [1990] ECR I–4585; and Case 314/89, [1991] ECR I–1647, [1993] 1 CMLR 171.

applicants by very reason of their non-marketing and conversion undertak-ings.[59] Of note in this context is the Court's rejection of argument that the 60 per cent rule was required in order to limit the increase in the Commu-nity reserve to 600,000 tonnes. Instead, it suggested an alternative means of finding allocations for the SLOM producers: if such allocations could not be found from the Community reserve without risk of disturbing the balance of the milk market, 'the fact remains that it would have been sufficient to reduce the reference quantities of the other producers proportionately by a corresponding amount'.[60] No specific mention was made of the interests of those other producers.[61]

Further, the Court elected not to follow the Opinion of the Advocate-General, who stated that 'given the impossibility of achieving absolute equality between returning and continuing producers, and as between the different categories of returning producers, the 60 per cent restriction must be regarded as proportionate and reasonable'. Central to his reason-ing was the fact that the 60 per cent rule corresponded with the general Community interest in curbing the dairy surplus — and in any event it did not remove a farmer's right to return to milk production.[62] The Court's decision to find the 60 per cent rule invalid, notwithstanding the conse-quences for the Community institutions in the administration of the market and notwithstanding the potential consequences for other producers, marks a further step in the development of the principle of protection of legitimate expectations. It may also be seen as a part of a continuum leading from *Mulder I* to *Mulder* v. *Council and Comission* ('*Mulder II*'), where the Court stated that the exclusion of the SLOM producers gave rise to a breach of a 'superior principle of law'.[63] At the same time it was decided in *Spagl* that the cut-off date was equally invalid, again for breach of the principle of protection of legitimate expectations. The theoretical possibil-ity of a cut-off date was accepted; but that actually imposed was unaccept-able, since again it specifically affected the applicant by very reason of his undertaking.[64]

In the third of these cases, *Rauh* v. *Hauptzollamt Nürnberg–Fürth*, the principle of protection of legitimate expectations once more provided the basis of the applicant's success.[65] His parents had entered into a five year

[59] See also Case 49/89, *Von Deetzen II*, [1991] ECR I–5119, [1994] 2 CMLR 487.
[60] Case 189/89, at I–4583 (ECR); Case 217/89, at I–4604 (ECR). For a similar willingness to hold to the general principles of Community law at the expense of practical and economic consequences, see, e.g., O'Reilly, J., 'Judicial Review and the Common Fisheries Policy in Community Law', in Curtin, D., and O'Keefe, D. (edd.), *Constitutional Adjudication in European Community and National Law: Essays for the Hon. Mr. Justice T. F. O'Higgins*, at 51–65.
[61] For consideration of their interests, see Council Reg. (EEC) 1639/91, [1991] OJ L150/35, Preamble.   [62] Case 189/89, at I–4570 (ECR).
[63] Case Joined Cases 104/89 and 37/90, [1992] ECR I–3061.
[64] Case 189/89, [1990] ECR I–4539.
[65] Case 314/89, [1991] ECR I–1647, [1993] 1 CMLR 171.

non-marketing scheme which expired on 21 December 1984.[66] The applicant subsequently took over the farm on 1 January 1985. The Court held that the right to SLOM 1 quota extended not only to the farmers who had entered into the non-marketing or conversion undertakings, but also to those who acquired their holdings by inheritance or in a similar manner following the expiration of those undertakings.

These judgments prompted the allocation of SLOM 2 quota under Council Regulation (EEC) 1639/91.[67] The 60 per cent rule was amended in the case of wholesale quota so as to provide for the deduction of a percentage representative of all the abatements applied to reference quantities, including in any case a basic reduction of 4.5 per cent. The basic reduction corresponded with the 4.5 per cent temporarily withdrawn under Council Regulation (EEC) 775/87[68] (as amended by Council Regulation (EEC) 3882/89[69]). The provisions were somewhat different with regard to direct sales quota. Again there was the deduction of a percentage representative of all the abatements applied to reference quantities; but, there having been no temporary withdrawal, no basic reduction was imposed. In *Kamp* v. *Hauptzollamt Wuppertal* the European Court was required to interpret this legislation; and, in particular, whether the basic reduction applicable to wholesale quota was to be aggregated with the representative percentage.[70] It confirmed that aggregation was indeed required; but that in calculating the representative percentage no account was to be taken of the basic reduction, since the latter was different in kind, being temporary rather than permanent.

The cut-off date was amended by permitting provisional allocations to producers whose non-marketing or conversion undertakings expired after 31 December 1982. As indicated, the European Court had accepted a cut-off date in principle; and the legislation justified the revised date on the grounds that, where a Member State's reference year was 1983, 'a producer who, with every opportunity to do so, had not resumed milk production between 1 January 1983 and 1 April 1984, had adequately demonstrated his wish to abandon milk production definitively for personal reasons not connected to the undertaking given or its consequences'.[71] Finally, the partial invalidity exposed in the decision of *Rauh* v. *Hauptzollamt Nürnberg–Fürth* was rectified by permitting applications by producers who, prior to 28 June 1989, had received the holding through an inheritance or similar means following the expiry of non-marketing or conversion undertakings entered into by the originator of the inheritance.

In addition, as with SLOM 1 quota, there were measures to ensure that

---

[66] For that reason, even prior to *Spagl* the cut-off date was satisfied.
[67] [1991] OJ L150/35.   [68] [1987] OJ L78/5.   [69] [1989] OJ L378/6.
[70] Case 21/92, [1994] ECR I–1619.
[71] Council Reg. (EEC) 1639/91, [1991] OJ L150/35, Preamble. The revised cut-off date survived challenge in Case 85/90, *Dowling* v. *Ireland*, [1992] ECR I–5305, [1993] 1 CMLR 288.

allocations were restricted to those who had the intent and capacity to resume milk production. For example, allocations were provisional, only becoming definitive if direct sales and/or deliveries reached 80 per cent of those provisional allocations within two years of 1 July 1991. Further, any sale or lease of the holding before 1 July 1994 resulted in the SLOM 2 quota being returned to the national reserve (proportionately on the basis of the fodder area if only part of the holding was sold or leased).

4.3.3. SLOM 3 allocations

The third round of SLOM allocations was made pursuant to two further decisions of the European Court, *Wehrs* v. *Hauptzollamt Lüneberg* and *Twijnstra* v. *Minister van Landbouw, Natuurbeheer en Visserij*.[72] In the former the applicant had already obtained a wholesale reference quantity as a result of production in his Member State's reference year. This allocation, according to the detailed provisions of Council Regulation (EEC) 764/89[73], precluded any grant of SLOM 1 quota in respect of additional land which he had purchased on 23 March 1984 subject to a conversion undertaking. Again the provisions were found invalid for breach of the principle of protection of legitimate expectations without need to address the other arguments before the Court; and again emphasis was laid on the fact that the applicant had been specifically affected by very reason of the undertaking under Council Regulation (EEC) 1078/77.[74]

In the latter case the applicant had sold part of his holding during the course of his non-marketing undertaking (which extended from 10 April 1980 to 9 April 1985). Since the purchaser agreed in writing to comply with the undertaking, the applicant was permitted to retain all the non-marketing premiums in accordance with the Community rules.[75] Having received all the premiums, he then claimed the full allocation of SLOM 1 quota when such became available. The European Court held, *inter alia*, that there would be a breach of the principle of protection of legitimate expectations if the applicant were to receive the full allocation. The purchaser was in the same position as the other SLOM producers, in that he could expect to use the land acquired for milk production on the expiry of the undertaking. As a result the allocation was to be divided proportionately between the applicant and the purchaser. Moreover, this was consistent with the imperative need to maintain the stability of the market, since the decision would not lead to a greater quantity of milk being delivered.

Both of these defects were remedied by Council Regulation (EEC) 2055/93; and, as with SLOM 1 quota and SLOM 2 quota, there were again

---

[72] Case 264/90, [1992] ECR I–6285; and Case 81/91, [1993] ECR I–2455.
[73] [1989] OJ L84/2.   [74] [1977] OJ L131/1.   [75] Ibid., Art.6(2).

restrictions affecting both entitlement and transfer.[76] For example, there was a measure to ensure that producers could increase production by the amount for which they applied; and, in the event of sales or leases of all or part of the holding before 1 October 1996, the SLOM 3 quota reverts to the national reserve in proportion to the area sold or leased.[77] Moreover, the leasing of SLOM 3 quota to other producers is not possible until 31 December 1997.[78] This being the third set of substantial amendments to accommodate the position of the SLOM producers, the degree of refinement was very considerable; and the resulting complexity has not escaped criticism.[79]

4.3.4. The resulting increase in reference quantities

Finally, the effect of the European Court's decisions was to require a yet further increase in the wholesale quota guaranteed total quantity of certain Member States. Again the United Kingdom was a substantial beneficiary, receiving as from the 1993/4 milk year some 84,675 tonnes, a 0.6 per cent increase.[80]

4.4. Further aspects of the litigation on SLOM allocations

Despite these notable successes, not all litigants were successful. On several occasions the European Court upheld the more restrictive approach favoured by the Council and Commission; and in this context may be discerned a growing concern on the part of all the Community institutions to prevent milk quota conferring an undeserved financial bonus. Instead, greater and greater emphasis is placed on the relationship between allocations of reference quantities and actual milk production.

Early and leading authority is provided by the case of *Von Deetzen II*, in which the applicant contested the validity of the restrictions on the transfer

---

[76] [1993] OJ L187/8. See also Comm. Reg. (EEC) 2562/93, [1993] OJ L235/18. For implementation in the UK, see the Dairy Produce Quotas (Amendment) Regs. 1994, S.I.1994 No.160.

[77] It is expressly provided that the holding comprises both the holding acquired and the other production units operated by the producer: Council Reg. (EEC) 2055/93, [1993] OJ L187/8, Art.4. For current implementation in the UK of this forfeiture to the national reserve, see the Dairy Produce Quotas Regs. 1994, S.I.1994 No.672, reg.8.

[78] Council Reg. (EEC) 2055/93, [1993] OJ L187/8, Art.4.

[79] See, e.g., Court of Auditors, *Special Report No.4/93 on the Implementation of the Quota System intended to control Milk Production*, [1994] OJ C12/1, at 4.10–11.

[80] Council Reg. (EEC) 1560/93, [1993] OJ L154/30. See also Federation of UK Milk Marketing Boards, *United Kingdom Dairy Facts and Figures: 1993 Edition* (Thames Ditton, Surrey, 1993), at Table 34. On this occasion, however, both France and Germany received greater amounts.

of SLOM 1 quota.[81] Of considerable significance is that, in sanctioning these restrictions, the Court rejected arguments based not only on the principle of non-discrimination and the right to property, but also on the principle of protection of legitimate expectations. With regard to the principle of protection of legitimate expectations, the Court affirmed that such producers could legitimately expect to be able to resume milk production under conditions which involved no discrimination. However, 'they could not thereby expect that a common organization of the market would confer on them a commercial advantage which did not derive from their occupational activity'.[82] SLOM quota was allocated to enable them to resume such occupational activity, not to dispose of for a windfall profit. Likewise there was no breach of the prohibition against discrimination: although the provisions specifically affected SLOM producers, they were justified by the need to prevent such producers using the allocations to derive a purely financial advantage. Finally, the right to property could not be invoked to cover the 'right to dispose, for profit, of an advantage, such as the reference quantities allocated within the framework of the common organization of a market, which does not derive from the assets or occupational activity of the person concerned'. Moreover, the right to property and other fundamental rights were not 'unfettered prerogatives but must be viewed in relation to their social function'.[83] Accordingly, the Court took the opportunity to express its disapproval of speculation in the value of milk quota and to reassert the link between allocations and occupational activity.[84] At the same time the judgment is very much in line with the Preamble to the implementing legislation, Council Regulation (EEC) 764/89; and, in particular, the clear statement that SLOM 1 quota was not intended to confer an 'undue advantage'.[85]

The provisions intended to ensure that applicants for SLOM quota had both the *bona fide* intent and capacity to resume milk production were also referred to the European Court; and in similar fashion the Court on the one hand showed a willingness to accommodate those who were genuinely striving to recommence their dairy activities, while on the other hand holding the line against those who might seek to realize a windfall profit. The legislation itself provided a high level of complexity; and this was exacer-

---

[81] Case 44/89, [1991] ECR I–5119, [1994] 2 CMLR 487. As indicated, if the holding were sold or leased prior to 31 Mar. 1992, the SLOM 1 quota was to be returned to the Community reserve (proportionately on the basis of the feed-crop area if only part of the holding was sold or leased).

[82] Case 44/89, [1991] ECR I–5119, [1994] 2 CMLR 487, at I–5155 (ECR) and 510 (CMLR).

[83] Ibid., at I–5156–7 (ECR) and 511 (CMLR). In this context the Court cited Case 265/87, *Hermann Schräder HS Kraftfutter GmbH & Co. KG* v. *Hauptzollamt Gronau*, [1989] ECR 2237 (a decision on the co-responsibility levy in the cereal sector).

[84] See, e.g., Case 2/92, *R.* v. *Ministry of Agriculture, Fisheries and Food*, ex p. *Bostock*, [1994] ECR I–955, [1994] 3 CMLR 547. For limitations on transfers more generally, see Ch.4.

[85] [1989] OJ L84/2.

bated by detailed factual matrices, some cases being heard a decade or more after the farmers concerned had entered into their non-marketing or conversion undertakings. As a result, changes in circumstances frequently rendered it impossible to resume milk production in exactly the manner employed prior to those undertakings.

As has been seen, for the provisional allocations of SLOM 1 quota to be made definitive, it was necessary for the SLOM producer to demonstrate that within two years of 29 March 1989 he had made direct sales and/or deliveries during the preceding twelve months amounting to at least 80 per cent of the provisional allocation. Further, he was obliged to show that he still operated in whole or part the same holding as that he operated when his premium application was approved. In *Maier* v. *Freistaat Bayern* the European Court upheld the latter of these criteria.[86] The farmer in question had entered into a conversion scheme, subsequently letting out his holding for a period of twenty years on 1 January 1987. His claim for SLOM 1 quota was precluded: he could not be said to be operating the same holding as that at the time of the premium application. Arguments based on the principle of protection of legitimate expectations and the prohibition of discrimination were all rejected, the Court (as in *Von Deetzen II*) being astute to prevent him realizing a windfall profit.

In *O'Brien* v. *Ireland* the applicant enjoyed more success;[87] but the judgment in his favour none the less contained reaffirmation of the approach adopted in *Von Deetzen II* and *Maier* v. *Freistaat Bayern*. He sought to render his provisional allocation definitive by production in part from the holding which had been subject to the non-marketing undertaking and in part from land taken on licence from his brother. This licence was but one aspect of a more complicated arrangement. In addition, the applicant took a lease of forty cows; and then the licensed land, the cows, and his own freehold land formed his capital contribution to a partnership between them.[88] The national court in Ireland, influenced by the restrictive nature of Council Regulation (EEC) 764/89, held that the only production to be taken into account was that from the holding which had been subject to the non-marketing undertaking (or so much of it as remained in the applicant's ownership).[89] However, this judgment was not followed by the European Court, which looked more generously at the detailed provisions. Under these the producer retained his entitlement provided that he had not abandoned the operation of all the holding subject to the undertaking. Accord-

---

[86] Case 236/90, [1992] ECR I–4483.

[87] Case 86/90, [1992] ECR I–6251, [1993] 1 CMLR 489.

[88] The European Court did not discuss in terms the validity of such joint venture arrangements as means of rendering provisional allocations definitive. On this aspect, see *R.* v. *Ministry of Agriculture, Fisheries and Food*, ex p. *St. Clere's Hall Farm*, [1995] 3 CMLR 125, post.

[89] [1991] 2 IR 387, [1990] ILRM 466.

ingly, in the absence of total abandonment, account could be taken of sales or deliveries from production units added to the original holding between the date of the undertaking and the date of the provisional allocation, it being sufficient that the producer operated only some part of the holding covered by the undertaking.[90] Emphasis was laid on the resumption of production itself rather than the land employed for that purpose. Thus, while this decision permitted the applicant some latitude, it did also provide the Court with an opportunity to re-emphasize the importance of the *activity* of milk production, consistent with its earlier judgments.

The same criteria were adopted by the United Kingdom court when called upon to consider the precise operation of such joint ventures. The applicants in *R.* v. *Ministry of Agriculture, Fisheries and Food*, ex parte *St. Clere's Hall Farm* had entered into partnerships for the purposes of rendering their SLOM 2 allocations definitive.[91] However, on the facts there seemed to be no significant pooling of assets; and the court was astute to prevent arrangements under which the recipient of a provisional SLOM allocation, not himself engaged in farming, entered into a partnership with an established producer with a view to rendering the SLOM allocation definitive and turning the resultant surplus to account. For that reason it was legitimate for the Ministry to withdraw the allocations.

Over and above this emphasis on occupational activity, the European Court has maintained a firm line on other restrictions imposed by the Community legislature. For example, it confirmed the validity of the rule under which SLOM 1 allocations were to be reduced by the amount of any reference quantities received in the 'special situations' as set out in Article 3(1) and (2) of Council Regulation (EEC) 857/84 and/or under the restructuring provisions in Article 4(1)(b) and (c) of the same regulation.[92] In the case of *R.* v. *Ministry of Agriculture, Fisheries and Food*, ex parte *Dent* it was claimed that an 'exceptional hardship' allocation made under the United Kingdom legislation implementing Article 4(1)(c) of Council Regulation (EEC) 857/84 did not fall to be deducted from the SLOM 1 award (Article 4(1)(c) relating to producers undertaking farming as their main occupation).[93] The claim was based upon close interpretation of Council Regulation (EEC) 764/89,[94] which provided for the deduction of a reference quantity obtained, *inter alia*, under Article 4(1)(b) *and* (c) of Council Regu-

---

[90] At the hearing the Commission had argued that the 'heart' of the holding must have remained unchanged between the date of the premium application and the period during which the provisional allocation was to be rendered definitive.

[91] [1995] 3 CMLR 125.

[92] [1984] OJ L90/13. Cf., Case 264/90, *Wehrs* v. *Hauptzollamt Lüneberg*, [1992] ECR I–6285 (where the anti-cumulation provisions did not prevent a producer who had received a reference quantity consequent upon production in his Member State's reference year from receiving also SLOM quota consequent upon the purchase of a farm subject to a conversion undertaking).

[93] Case 84/90, [1992] ECR I–2009, [1992] 2 CMLR 597.   [94] [1989] OJ L84/2.

lation (EEC) 857/84. The applicants alleged that they should not be subject to the deduction, since they had only received a reference quantity under Article 4(1)(c): indeed, Article 4(1)(b) (which related to certain development plans) had not even been implemented in the United Kingdom. The claim was rejected, since otherwise unjust enrichment would flow from two reference quantities being allocated in respect of the same production. While this decision may be narrow in ambit, it is again illustrative of the complexity of the legislation and of the scope for references to the European Court.

A final example is provided by the case of *Dowling* v. *Ireland*.[95] The legislative background was somewhat distinctive, the national court and European Court being required to address the interaction of the reference year provisions under the original legislation and the amended cut-off date under the SLOM legislation.[96] However, again a strict approach was adopted. The applicant had been unfortunate. His non-marketing undertaking had expired in November 1982, prior even to the amended cut-off date, but he had been unable to resume farming until 1984 as a result of ill health. Thus, the combination of the non-marketing undertaking and health difficulties prevented milk production throughout the period 1981–3; and consequently he had received no reference quantity whatsoever. The European Court held that he was indeed precluded from an allocation. On the one hand, the alternative reference year provisions were of no assistance. His incapacity might entitle him to choose an alternative reference year to that applicable in Ireland (i.e., 1983),[97] but his choice was restricted to 1981 or 1982, during neither of which he had representative production by reason of his non-marketing undertaking.[98] On the other hand, he was not entitled to a SLOM quota allocation, the amended cut-off provisions excluding those whose undertakings expired before 1 January 1983. Furthermore, the rules so interpreted did not breach the principle of protection of legitimate expectations: the reason for the applicant's failure to market milk between the expiration of his undertaking and the implementation of the milk quota system was unconnected with the undertaking. Nor was there an infringement of the prohibition of discrimination: the rules were objectively justified in the interests of legal certainty and the effectiveness of the system.[99]

---

[95] Case 85/90, [1992] ECR I–5305, [1993] 1 CMLR 288. For the decision of the national court, see [1991] 2 IR 379, [1991] 2 CMLR 193.

[96] I.e., as amended by Council Reg. (EEC) 1639/91, [1991] OJ L150/35.

[97] See Council Reg. (EEC) 857/84, [1984] OJ L90/13, Art.3(3), as expanded by Comm. Reg. (EEC) 1546/88, [1988] OJ L139/12, Art.3 (re-enacting Comm. Reg. (EEC) 1371/84, [1984] OJ L132/11, Art.3).

[98] In support of this approach the Court cited Case 84/87, *Erpelding* v. *Secrétaire d'État à l'Agriculture et à la Viticulture*, [1988] ECR 2647, [1989] 3 CMLR 493.

[99] The Court again cited Case 84/87, *Erpelding* v. *Secrétaire d'État à l'Agriculture et à la Viticulture*, [1988] ECR 2647, [1989] 3 CMLR 493.

4.5. Compensation claims against the Community institutions

Not content with their successes in securing allocations of reference quantities, SLOM producers claimed that the partial invalidities exposed in the Community legislation were sufficient to give rise to liability under Article 215(2) of the EC Treaty which provides that: 'In the case of non-contractual liability, the Community shall, in accordance with the general principles common to the laws of the Member States, make good any damage caused by its institutions or by its servants in the performance of their duties'.[100] When the issue came before the European Court in *Mulder II*, the judgment then given marked a significant development in Community jurisprudence. Not least it clarified the circumstances in which such liability would be incurred.[101]

Four aspects of the judgment may be highlighted. First, the Court dismissed argument that damage was attributable to the national authorities rather than the Community. The ground relied upon by the Council and the Commission was that it had been open to the national authorities to make use of the various provisions in Council Regulation (EEC) 857/84, as amended, which permitted allocations in special situations or to complete restructuring and the reallocation of unutilized reference quantities.[102] In the Court's view this ground could not be sustained when no obligation to implement such provisions had been imposed on the national authorities. Moreover, the provisions were not designed to address the position of the SLOM producers.[103]

Secondly, as indicated, the court laid down the circumstances in which the Community would incur non-contractual liability under Article 215(2). Three criteria were identified.[104] First, there must be 'a sufficiently serious breach of a superior rule of law for the protection of the individual'. Secondly, to give rise to such a breach, the Community institution in question

---

[100] For fuller discussion of the non-contractual liability of the Community, see, e.g., Schermers, H.G., Heukels, T., and Mead, P. (edd.), *Non-contractual Liability of the European Communities* (Europa Instituut, University of Leiden, Nijhoff, 1988), *passim*; Schockweiler, F., 'Le régime de la responsibilité extra-contractuelle du fait d'actes juridiques dans la Communauté européenne', (1990) 26(1) R.T.D.E. 27–74; Wils, W., 'Concurrent Liability of the Community and a Member State', (1992) 17 E.L.Rev. 191–206; Wyatt, D. and Dashwood, A., *European Community Law* (3rd edn., Sweet and Maxwell, London, 1993), at 156–62; Hartley, T. C., *The Foundations of European Community Law* (3rd edn., Clarendon Press, Oxford, 1994), at 467–507; Lasok, D., *Law and Institutions of the European Union*, (6th edn., Butterworths, London, 1994), at 48–56; and Steiner, J., *Textbook on EC Law* (4th edn., Blackstones Press, London, 1994), at 388–402.

[101] Joined Cases 104/89 and 37/90, [1992] ECR I–3061. See also Heukels, T., (1993) 30 C.M.L.Rev. 368–86; and Brown, L. N. and Kennedy, T., *The Court of Justice of the European Communities*, at 166–8.

[102] I.e., Council Reg. (EEC) 857/84, [1984] OJ L90/13, Arts.3, 4, and 4a.

[103] See also Case 175/84, *Krohn and Co. Import-Export GmbH & Co. KG* v. *Commission*, [1986] ECR 753, [1987] 1 CMLR 745.

[104] Joined Cases 104/89 and 37/90, [1992] ECR I–3061, at 3131–2.

must have 'manifestly and gravely disregarded the limits on the exercise of its powers' — this being particularly so in the context of the Common Agricultural Policy, where the Community institutions enjoyed broad discretion.[105] Thirdly, the damage pleaded must 'go beyond the bounds of the normal economic risks inherent in the activities in the sector concerned'.[106]

The partial invalidity exposed in Council Regulation (EEC) 857/84 satisfied these criteria.[107] In accordance with the decisions in *Mulder I* and *Von Deetzen I*, the regulations breached the principle of protection of legitimate expectations, which constituted a general and superior principle of Community law for the protection of the individual.[108] This breach was 'sufficiently serious' in that the legislature had 'failed completely, without invoking any higher public interest, to take account of the specific situation of a clearly defined group of economic agents' i.e., the SLOM producers. Finally, total and complete exclusion from the allocation of a reference quantity, which precluded the resumption of milk production, could not 'be regarded as foreseeable or as falling within the bounds of the normal economic risks inherent in the activities of a milk producer'.[109]

By contrast, Council Regulation (EEC) 764/89 gave rise to no such liability.[110] Although the 60 per cent rule breached the principle of protection of legitimate expections, the breach was not sufficiently serious for the purposes of Article 215(2), in that the farmers could resume milk production (albeit at a reduced level). Accordingly, it could not be argued that the Council had failed to take them into account. Further, the legislature had made an economic policy choice, based, on the one hand, on 'the overriding necessity of not jeopardizing the fragile stability that currently obtains in the milk products sector' (as set out in the Preamble to Council Regulation (EEC) 764/89); and, on the other hand, on the need to balance the interests of the SLOM producers and other producers subject to the milk quota system. In this context, the increase in the Community reserve by 600,000

[105] In support of these two criteria the Court cited Joined Cases 83 and 94/76, 4, 15, and 40/77, *Bayerische HNL Vermehrungsbetriebe GmbH and Co. KG* v. *Council and Commission*, [1978] ECR 1209, [1978] 3 CMLR 566.

[106] In support of this criterion the Court cited Case 238/78, *Ireks-Arkady GmbH* v. *Council and Commission*, [1979] ECR 2955; Joined Cases 241, 242, and 245–250/78, *DGV* v. *Council and Commission*, [1979] ECR 3017; Joined Cases 261 and 262/78, *Interquell Stärke-Chemie GmbH & Co. KG* v. *Council and Commission*, [1979] ECR 3045; and Joined Cases 64 and 113/76, 167 and 239/78, 27, 28, and 45/79, *Dumortier Frères SA* v. *Council*, [1979] ECR 3091.

[107] [1984] OJ L90/13, as expanded by Comm. Reg. (EEC) 1371/84, [1984] OJ 132/11.

[108] Case 120/86, [1988] ECR 2321, [1989] 2 CMLR 1; and Case 170/86, [1988] ECR 2355, [1989] 2 CMLR 327.

[109] Joined Cases 104/89 and 37/90, [1992] ECR I–3061, at I–3132–3. See also Case 152/88, *Sofrimport SARL* v. *Commission*, [1990] ECR I–2477, [1990] 3 CMLR 80 (the applicants likewise succeeded under Art.215(2) of the EC Treaty, where the Commission had failed completely to take their interests into account).

[110] [1989] OJ L84/2.

tonnes was the greatest amount consistent with the aims of the milk quota system; and 600,000 tonnes only realized some 60 per cent of the foreseeable demand for SLOM 1 quota. Thus, in line with the higher public interest, the Council had not manifestly and gravely disregarded the limits of its discretionary power. It is of note that, as already indicated, the Advocate-General in *Spagl* v. *Hauptzollamt Rosenheim* and *Pastätter* v. *Hauptzollamt Bad Reichenhall* had been prepared to accept the validity of the 60 per cent rule for similar reasons.[111] Besides, the correct invocation of higher public interest may be seen as a key determining factor.

In laying down and applying these criteria, the Court largely followed the tests derived from *Aktien-Zucherfabrik Schöppenstedt* v. *Council* (not expressly mentioned in the judgment)[112] and *Bayerische HNL Vermehrungsbetriebe GmbH & Co. KG* v. *Council and Commission*[113] — tests clearly stated by the Advocate-General in *Mulder II* itself.[114] However, in one respect the Court would appear to have diverged from earlier orthodoxy. According to the judgment, it seemed sufficient for the Court that the applicants be 'clearly defined'. This marked a departure from the more traditional view that the group also be limited in extent — a requirement repeated by the Advocate-General in *Mulder II*. That having been said, in the view of the Advocate-General this criterion would still have been satisfied by the SLOM producers, since the expression 'limited' did not impose an absolute numerical ceiling on persons adversely affected.[115]

Thirdly, the Court provided remarkably detailed guidelines as to the assessment of compensation. In the absence of specific circumstances justifying an alternative form of assessment, account was to be taken of the loss of earnings suffered by the applicants. This loss consisted of the difference between, 'on the one hand, the income which the applicants would have obtained in the normal course of events from the milk deliveries which they would have made . . .' during the period from 2 April 1984 (when milk quotas were introduced) to 29 March 1989 (when SLOM 1 allocations were authorized by Council Regulation (EEC) 764/89);[116] and, 'on the other hand, the income which they actually obtained from milk deliveries made during that period in the absence of any reference quantity, plus any income which they obtained, or could have obtained, during that period from any replacement activities'.[117]

[111] Case 189/89, [1990] ECR I–4539; and Case 217/89, [1990] ECR I–4585.
[112] Case 5/71, [1971] ECR 975.
[113] Joined Cases 83 and 94/76, 4, 15 and 40/77, [1978] ECR 1209, [1978] 3 CMLR 566.
[114] Joined Cases 104/89 and 37/90, [1992] ECR I–3061. See also Case 152/88, *Sofrimport SARL* v. *Commission*, [1990] ECR I–2477, [1990] 3 CMLR 80.
[115] As seen, some 13,187 applications were made for SLOM 1 quota alone: Written Answer given by Mr MacSharry on behalf of the Commission to a question from Mr Hume: [1990] OJ C93/26. See also Heukels, T., (1993) 30 C.M.L.Rev. 368–86, at 380–3; and Steiner, J., *Textbook on EC Law*, at 392–6.
[116] [1989] OJ L84/2.   [117] Joined Cases 104/89 and 37/90, [1992] ECR I–3061, at I–3135.

Specific aspects of this calculation received elaboration. For the purposes of determining the reference quantities to which the applicants would have been entitled between 2 April 1984 and 29 March 1989, it might first be necessary (as with SLOM 1 allocations themselves) to ascertain the quantity of milk upon which the non-marketing or conversion premium was based. This quantity would then require adjustment to bring it into line with the reference quantities of farmers who had remained in production.[118] The next stage was to calculate the income from hypothetical deliveries corresponding to the applicants' reference quantities as so determined. The basis of this calculation would be the profitability of a farm representative of the type of farm run by each of the applicants. However, account was to be taken of the reduced profitability generally shown when milk production was commenced. Finally, the Court emphasized that from the resulting figure should be deducted not only the actual amount of income derived from replacement activities, but also the amount of any income which could have been obtained.

In this connection two points may be emphasized. First, the calculation was demonstrably complex. In part this may be seen as a product of the numerous factors which had come into play since the commencement of the non-marketing and conversion undertakings. At the same time, the position was not eased by the Court's willingness to tailor the assessment of compensation to address the circumstances of individual producers. Secondly, the Court expressly accepted the general principle of mitigation common to the legal systems of Member States. Accordingly, 'the injured party must show reasonable diligence in limiting the extent of his loss or risk having to bear the damage himself'.[119]

Fourthly, and finally, the Court awarded interest as from judgment.[120] In Case 104/89, *Mulder* v. *Council and Commission*, the rate was fixed at 8 per cent per annum, while in Case 37/90, *Heinemann* v. *Council and Commission*, the rate was only 7 per cent per annum, the amount for which the applicant had applied.[121] No other reasoning was given — and the Court did

---

[118] More specifically, the applicants would receive the blanket 1 per cent increase under Council Reg. (EEC) 857/84, [1984] OJ L90/13, Art.2(1) (to prevent them suffering undue disadvantage) — but they would also be subject to a reduction representative of the rates applicable to farmers who had continued in production. However, in the latter regard, no account was to be taken of the percentage weighting imposed by Art.2(2) where a Member State chose an alternative reference year to the 1981 calendar year: that percentage weighting was designed to offset the benefits of increased production in 1982 and 1983, to which the applicants did not contribute. Further, in establishing the representative rate of reduction, account was to be taken of compensation payable (as for the reference quantities temporarily withdrawn under Council Reg. (EEC) 775/87, [1987] OJ L78/5). On these points generally, cf., Case 21/92, *Kamp* v. *Hauptzollamt Wuppertal*, [1994] ECR I–1619.

[119] Joined Cases 104/89 and 37/90, [1994] ECR I–3061, at I–3136–7.

[120] See also Joined Cases 64 and 113/76, 167 and 239/78, 27, 28, and 45/79, *Dumortier Frères SA* v. *Council*, [1979] ECR 3091.

[121] Cf., e.g., Case 256/81, *Paul's Agriculture Ltd.* v. *Council and Commission*, [1983] ECR

not follow the view expressed by the Advocate-General to the effect that the guideline should be prevailing interest rates at the date of judgment in the Member States where the applicants worked and in which they would therefore normally use or invest the compensation.[122]

Following the judgment in *Mulder II*, and to ensure orderly treatment of the plethora of claims, the Council and Commission published a Communication regarding entitlement to compensation.[123] This contained full recognition of the non-contractual liability of the Community as against all producers in a comparable situation to that of the applicants in *Mulder II*. It also had the broad effect of suspending the operation of Article 43 of the Statute of the Court of Justice (under which proceedings against the Community arising from non-contractual liability are barred after a period of five years from the occurence of the event giving rise to such liability). Provided entitlement was not already time-barred by the date of the Communication (5 August 1992), or on an earlier date when the application had been made, then the Community institutions undertook not to rely on Article 43 for the purpose of defeating any such actions.

Detailed rules emerged approximately one year later; and these illustrate graphically the practical difficulties encountered.[124] Indeed, the Preamble to Council Regulation (EEC) 2187/93 recited that 'the sheer number of those potentially eligible makes it impossible to take each case into account on an individual basis'. The solution adopted was a flat-rate offer, to be accepted within two months of receipt in full and final settlement. If not accepted within that time, it would cease to be binding in the Community institutions concerned. To allow for the absence of specific consideration of individual claims, factors chosen to calculate the compensation were expressed to 'operate as a general rule to the benefit of the applicants'. In particular, the overall bluntness of the offer was mitigated by introducing different rates of compensation according to the year concerned and the size of the holding. The time-bar provisions as promulgated in the Communication of 5 August 1992 were enacted by Article 8 of the regulation. Council Regulation (EEC) 2187/93 was itself supplemented by Commission Regulation (EEC) 2648/93, making available a voluntary contribution by the Community towards agents' costs, again at a flat-rate.[125]

Not even these regulations have remained immune from attack, the

---

1707, [1983] 3 CMLR 176 (6%); and Case 152/88, Sofrimport *SARL* v. *Commission*, [1990] ECR I–2477, [1990] 3 CMLR 80 (8%).

[122] For the possibility of discriminatory treatment arising from different levels of interest in different Member States, see Heukels, T., (1993) 30 C.M.L.Rev. 368–86, at 385.

[123] [1992] OJ C198/4.

[124] Council Reg. (EEC) 2187/93, [1993] OJ L196/6.

[125] [1993] OJ L243/1.

effectiveness of the time-bar provisions being challenged by both United Kingdom and Irish producers.[126] Again the scale of the litigation is substantial, with 247 applicants in the Irish case alone. Indeed, in order to reduce the burden on the European Court, it had been decided as early as June 1993 to transfer, *inter alia*, compensation claims to the Court of First Instance.[127] However, the applicants faced a formidable difficulty. The main application contesting the validity of the time-bar provisions would not be heard until long after the deadline for accepting the flat-rate compensation — yet to accept that compensation would preclude any further claim against the Community institutions.[128] For that reason the producers concerned made an application for interim measures, seeking an order that they could accept the flat-rate offer while continuing with their challenge to the time-bar provisions.[129] Not least, it was argued that serious and irreparable damage would be suffered in the absence of such interim measures: if the flat-rate offer were accepted, any claim against the Community institutions for the full amount of the damage would have to be abandoned; and, if it were rejected, pressing financial circumstances rendered it impracticable to await the outcome of the main proceedings. The Court of First Instance refused to grant the application, holding that acceptance of the flat-rate offer did not necessarily lead to the loss of all rights to more extensive compensation. In this regard it took note of a declaration by the Council and Commission that, if others persevered with the main proceedings and the time-bar provisions were subsequently found to be invalid, then a 'new situation' would have arisen as to the period for which compensation was payable; and the applicants could benefit from that new situation. However, in such circumstances, the basis of compensation might be revised in favour of, for example, actual losses.

5. CONCLUSION

The success enjoyed by the applicants in the SLOM litigation stands as a tribute to the ingenuity of both the producers themselves and their advisers. At the same time the various judgments betray a willingness on the part of

---

[126] The main actions by UK farmers are Cases T–278/93 R and T–555/93 R, *Jones* v. *Council and Commission*; and the Irish action is Case T–541/93 R, *McCutcheon* v. *Council*. See also, e.g., Moss, J. R., Davis, N., and Saunders, J. 'Milk Quota Update: Autumn 1993', (1993) *Bulletin of the Agricultural Law Association*, Issue 11, 8–10, at 9.

[127] Decision of the Member States of 8 June 1993, [1993] OJ L144/21. See also Brown, L. N. and Kennedy, T., *The Court of Justice of the European Communities*, at 75–6 and 167–8.

[128] Council Reg. (EEC) 2187/93, [1993] OJ 196/6, Art.14.

[129] Joined Cases T–278/93 R and T–555/93 R, T–280/93 R, and T–541/93 R, *Jones* v. *Council and Commission*, [1994] ECR II–11. See also Case T–554/93 R, *Abbott Trust* v. *Council and Commission*, [1994] ECR II–1.

the European Court to hold to the general principles of Community law notwithstanding the discomforture which may be caused to the Council and Commission. In this context may be highlighted the cases of *Spagl* v. *Hauptzollamt Rosenheim* and *Pastätter* v. *Hauptzollamt Bad Reichenhall*, where the 60 per cent rule was held invalid — despite the consequence that this could lead to reductions in the reference quantities of continuing producers or an increase in the Community reserve on a scale incompatible with the objective of the milk quota system.[130]

While recognizing that the different decisions are based on particular sets of complex facts, some consistent strands may be detected. First, the cases have helped to establish the principle of protection of legitimate expectations as one of the superior general principles of Community law — and one with broad effect. In particular, it has emerged as an important factor that the applicants were encouraged by the Community institutions to adopt a form of action in furtherance of Community policy and the general interest. In such circumstances, the legislature cannot lightly impose specific restrictions on individuals who were complying with Community initiatives.[131]

Secondly, there would seem to be a contrast between total exclusion from an area of economic activity and mere restrictions.[132] In *Mulder I* and *Von Deetzen I* the European Court emphasized that the applicant was effectively barred from resuming milk production — a state of affairs entirely different in scale to the more limited cuts upon production suffered by continuing producers.[133] Again in *Mulder II* one of the criteria for the Community's liability was satisfied by the fact that the legislature had completely failed to take account of the SLOM producers' situation, without invoking any higher public interest. This manifest and grave disregard of the limits of its discretionary power amounted to a serious breach of a superior rule of law. Further, such total exclusion could not be considered foreseeable; and the applicants had suffered damage beyond the normal economic risks in the milk sector.[134] The 60 per cent rule called for a more fine distinction. While in *Spagl* v. *Hauptzollamt Rosenheim* and *Pastätter* v. *Hauptzollamt Bad Reichenhall* the rule was held invalid for breach of the principle of protection of legitimate expectations, the decision was

[130] Case 189/89, [1990] ECR I–4539; and Case 217/89, [1990] ECR I–4585.
[131] See also Case 177/90, *Kühn* v. *Landwirtschaftskammer Weser–Ems*, [1992] ECR I–35, [1992] 2 CMLR 242. In the course of its judgment the Court stated that 'the principle of the protection of legitimate expectations may be invoked as against Community rules, only to the extent that the Community itself has previously created a situation which can give rise to a legitimate expectation': at I–63 (ECR) and 259 (CMLR).
[132] This approach was found earlier in Case 44/79. *Hauer* v. *Rheinland-Pfalz*, [1979] ECR 3727, [1980] 3 CMLR 42.
[133] Case 120/86, [1988] ECR 2321, [1989] 2 CMLR 1; and Case 170/86, [1988] ECR 2355, [1989] 2 CMLR 327.
[134] Joined Cases 104/89 and 37/90, [1992] ECR I–3061.

contrary to the Opinion of the Advocate-General;[135] and in *Mulder II* the European Court did not see fit to award compensation under that head.[136] During the course of rejecting the claim, emphasis was laid on the fact that the breach was insufficiently serious for the purposes of founding liability under Article 215(2) of the EC Treaty; and that the Council had taken account of a higher public interest, without gravely and manifestly disregarding the limits of its discretionary power in administering the Common Agricultural Policy.

Moreover, a distinction could be drawn between the circumstances of the SLOM producers, who simply were not covered by the implementing legislation, and the circumstances of those producers who were covered by the implementing legislation but, failing to satisfy the requisite criteria, received no allocation. Examples of producers in this second category are furnished by the applicants in *Erpelding* v. *Secrétaire d'État à l'Agriculture et à la Viticulture* and *Leukhardt* v. *Hauptzollamt Reutlingen*.[137] In this context the case of *Dowling* v. *Ireland* may be considered particularly illustrative, in that it addressed the position after the amendments had been enacted to make the SLOM allocations available.[138] Although the applicant attracted the sympathy of the Court, he received no reference quantity. On the one hand, he failed to qualify under the legislation as originally enacted and, on the other hand, he failed to qualify under the legislation as amended to remedy the invalidities exposed. Further, a point of critical importance, once the amendments had been enacted he could no longer claim that the legislature had failed to take account of his situation.

Thirdly, the Community institutions showed a clear dislike of speculation in the market value of milk quota. This dislike was manifested in, for example, the legislative requirement that SLOM producers should demonstrate the *bona fide* intent and capacity to recommence milk production;[139] and in the antipathy of the European Court towards farmers who saw in SLOM quota the opportunity to derive a windfall profit when the purpose of the allocation was to permit resumption of their occupational activity. This approach has been continued through into the more recent regulations. For example, Council Regulation (EEC) 2187/93 stipulates that full compensation is dependent upon the applicants having achieved a level of production sufficient to render their provisional allocations of SLOM quota definitive.[140] Indeed, the Preamble to that regulation may be seen as a clear

---

[135] Case 189/89, [1990] ECR I–4539; and Case 217/89, [1990] ECR I–4585.

[136] Joined Cases 104/89 and 37/90, [1992] ECR I–3061.

[137] Case 84/87, [1988] ECR 2647, [1989] 3 CMLR 493; and Case 113/88, [1989] ECR 1991, [1991] 1 CMLR 298.

[138] Case 85/90, [1992] ECR I–5305, [1993] 1 CMLR 288.

[139] Council Reg. (EEC) 764/89, [1989] OJ L84/2; and Council Reg. (EEC) 1639/91, [1991] L150/35.

[140] [1993] OJ L196/6.

exposition of the attitude of the Community institutions: 'a direct link should be drawn between the actual resumption of milk production in full compliance with Community provisions which permitted such resumption and the existence of an injury consisting in the fact of not having been able to resume milk production in good time, despite the wishes of the person concerned'; and 'in the same spirit and to ensure that a producer has not resumed activity with the sole aim of speculating on the supposed asset value of the reference quantity allocated to him, both the broad entitlement to compensation and the amount of that compensation should be subject to certain conditions'.

3

Reform and the Current Legislation

1. THE REASONS FOR REFORM

1.1. General

This series of further allocations would alone have been sufficient to place strain on the milk quota system. However, as the time arrived for the Community to decide whether or not to retain quotas beyond 31 March 1992, several other areas of persistent difficulty could be identified; and the need for reform became imperative.[1]

Of these other difficulties, two examples may be given. First, in certain Member States there was only fragmentary application of the milk quota system. During the course of its audit conducted in 1992, the Court of Auditors was disturbed to discover that Greece, Italy, and Spain were yet to complete the allocation of reference quantities; and the Commission received unqualified criticism for failing to take all the necessary actions to remedy this deficiency.[2] In particular, the Commission had obtained only one judgment from the European Court for the purposes of securing compliance with the implementing legislation.[3] Besides, on a more general level, this enforcement procedure under Article 169 of the EC Treaty suffered until recently from the lack of effective sanction in the event of non-compliance with a European Court judgment.[4]

Secondly, various amendments to the Community and national legislation had combined over the years to reduce the dissuasive effect of the milk quota system. Being technical and detailed in nature, their importance may easily escape notice. However, certain aspects may be highlighted: first, the increased use of the offsetting mechanism (leading to reduced liability on the part of individual producers); secondly, conversion between wholesale quota and direct sales quota; and, thirdly, butterfat adjustments. Other

---

[1] See, e.g., Court of Auditors, *Special Report No.4/93 on the Implementation of the Quota System intended to control Milk Production*, [1994] OJ C12/1, at 4.1–5.21.

[2] Ibid., at 4.6, 4.55, and 5.15.   [3] Case 394/85, *Commission v. Italy*, [1987] ECR 2741.

[4] For a full discussion of enforcement by the Commission under Art.169 of the EC Treaty, see, e.g., Wyatt, D. and Dashwood, A., *European Community Law*, at 109–20; and Brown, L. N. and Kennedy, T., *The Court of Justice of the European Communities*, at 105–13. Following the Maastricht Treaty, the Court as a last resort may impose a lump sum or penalty payment on Member States which fail to comply with a judgment: see now Art.171(2) of the EC Treaty, as inserted by the Maastricht Treaty.

aspects included delay in collecting the superlevy; and the judicious use of permanent and temporary transfers of quota as a means of escaping or limiting liability (which will be considered later in the context of quota transfers generally).[5]

1.2. Specific areas of difficulty

1.2.1. Offsetting

With regard to offsetting, it has already been seen that the original legislation permitted producers operating under Formula B to offset at the level of the purchaser or dairy.[6] Notwithstanding that individual producers made deliveries to a purchaser in excess of their reference quantities, there would be no liability to pay the superlevy provided that other producers affiliated to the same purchaser had sufficient unutilized reference quantities to cover the excess. To take account of this facility, the original legislation also provided for a higher rate of levy than that applied under Formula A or in the case of direct sales quota. However, the scope for offsetting was dramatically increased by Council Regulation (EEC) 590/85,[7] which inserted a new Article 4a into Council Regulation (EEC) 857/84.[8] Under Article 4a, which applied with retroactive effect for the 1984/5 milk year, Member States were authorized to allocate the unutilized reference quantities of producers or purchasers to, first, producers or purchasers in the same region and, if necessary, to producers or purchasers in other regions. The provision applied to both direct sales and wholesale quota.[9] Accordingly, the combination of the original legislation and the amendment effected by Council Regulation (EEC) 590/85 could ensure that individual producers enjoyed much less than full responsibility for their own production.[10] At the time enacted as a temporary measure to assist gradual adaption to the milk quota system, such offsetting nevertheless became an

---

[5] See Ch.4. For a full survey of all these aspects, see Court of Auditors, *Special Report No.4/ 93 on the Implementation of the Quota System intended to control Milk Production*, [1994] OJ C12/1, at 4.16–50.

[6] See, in particular, Case 61/87, *Thevenot* v. *Centrale Laitière de Franche–Comté*, [1988] ECR 2375, [1989] 3 CMLR 389.

[7] [1985] OJ L68/1.   [8] [1984] OJ L90/13.

[9] For implementation in the UK, see the Dairy Produce Quotas (Amendment) Regs. 1985, S.I.1985 No.509 (a new reg.9D and Sch.8 being inserted into the Dairy Produce Quotas Regs. 1984, S.I.1984 No.1047). On offsetting at the level of the purchaser and Art.4a generally, see, e.g., Eurostat 1990, Vol.1, at 17–21; and Gehrke, H., *The Implementation of the EC Milk Quota Regulations in British, French and German Law*, at 50–3.

[10] The amendment effected under Art.4a proved significant in Joined Cases 267–285/88, *Wuidart* v. *Laiterie coopérative eupenoise*, [1990] ECR I–435. As already seen, it was argued that Formula A producers suffered discrimination, in that they did not enjoy the ability to

established factor in the calculation of the superlevy.[11] Indeed, Council Regulation (EEC) 1110/88 recited that, 'given its flexibility, this arrangement should be retained, in view of certain situations, as long as the said levy is chargeable'.[12]

The full effect of this discretion afforded Member States is not easy to overestimate. A clear example is provided by the fact that, for the 1985/6 milk year, the actual rate of superlevy paid by Formula B producers in England and Wales was a mere 0.8 per cent of the target price for milk (the full rate at that date being 100 per cent of the target price).[13] An important factor in the equation was a surprising amount of underproduction to be exploited by those who exceeded their reference quantities. Thus, in England and Wales for the 1986/7 milk year there were some 365 million litres of unutilized wholesale quota (approximately 3 per cent of the national guaranteed total quantity), as opposed to some 490 million litres of individual excess production (approximately 4 per cent of the national guaranteed total quantity).[14]

Further amendments combined to reduce the deterrent effect of the superlevy. Thus, Council Regulation (EEC) 774/87 stated that it would be charged on all quantities in excess of individual reference quantities only after making 'any corrections';[15] and Council Regulation (EEC) 764/89 (in addition to allocating SLOM 1 quota) provided that for the 1984/5 and 1985/6 milk years the superlevy would only be payable where production exceeded the national quaranteed total quantity for wholesale and/or direct sales quota.[16]

That having been said, as from the 1987/8 milk year Member States operating Formula B were authorized to target the superlevy against those producers who had most exceeded their individual reference quantities.[17] Under the general rule, if deliveries were made above a purchaser's reference quantity, the purchaser was to recoup the burden of the superlevy from those producers who had contributed to the excess after allocating

offset at the level of the purchaser — and yet were obliged to pay the superlevy at a rate only 25% less than producers under Formula B. A factor influencing the European Court's dismissal of the claim was that, although Formula A producers could not offset at the level of the purchaser, there was a wider opportunity to do so under Art.4a.

[11] See Council Reg. (EEC) 1305/85, [1985] OJ L137/12; Council Reg. (EEC) 774/87, [1987] OJ L78/3; and Council Reg. (EEC) 1110/88, [1988] OJ L110/28.

[12] [1988] OJ L110/28.

[13] Court of Auditors, *Special Report No.2/87 on the Quota/Additional Levy System in the Milk Sector*, [1987] OJ C266/1, at 3.18 and Table 5. For an example of the method of calculation, see ibid., Annex.

[14] Eurostat 1990, Vol.1, at 17. Among the reasons suggested for this phenomenon were weather conditions, cattle diseases, and adaption to changes in reference quantities.

[15] [1987] OJ L78/3.   [16] [1989] OJ L84/2.

[17] Council Reg. (EEC) 773/87, [1987] OJ L78/1. See also Council Reg. (EEC) 744/88, [1988] OJ L78/1.

unutilized reference quantities in proportion to their individual reference quantities.[18] At the same time, even if the purchaser's reference quantity was not exceeded, Member States had the option to impose the superlevy in its entirety on all producers who had exceeded their individual reference quantities by at least 10 per cent or at least 20,000 kg.[19]

1.2.2. Conversion between wholesale quota and direct sales quota

Secondly, enhanced flexibility was conferred by the authorization of conversion between wholesale and direct sales quota. Such flexibility had been advocated from the very inception of the milk quota system;[20] and at the level of the producer it was implemented under Council Regulation (EEC) 590/85,[21] which inserted a new Article 6a into Council Regulation (EEC) 857/84.[22] However, the original Article 6a was somewhat restrictive: conversion was only applicable for the currency of a milk year; requests would only be entertained from producers who had both wholesale and direct sales reference quantities; and the producers were to be in a position such that it was necessary to adapt to changes in their marketing requirements.[23] That having been said, the European Court subsequently confirmed that it was sufficient for producers to hold both wholesale and direct sales reference quantities: it was not necessary for them actually to be making both wholesale deliveries and direct sales during the milk year of conversion.[24] Article 6a was soon accompanied by further scope for conversion, on this occasion at the level of the Member State rather than the producer. Council Regulation (EEC) 1298/85 rendered it possible for Member States to effect definitive adjustments of their guaranteed total quantities for wholesale and direct sales quota on the basis of objective statistical data and the structural development of the respective levels of production.[25]

---

[18] Member States were permitted to give priority to certain certain producers, a derogation not enacted in the UK.

[19] For the operation of the 'threshold system' applicable to wholesale quota as implemented in the UK, see Milk Marketing Board, *Five Years of Milk Quotas: a Progress Report*, at App. 6. While the purchaser's liability remained unchanged, the superlevy was payable by individual producers at the full target price for milk — but only by those who had most exceeded their reference quantity. The calculation was made as follows: first, all producers were ranked in descending order, according to the percentage by which they had exceeded their individual reference quantity; then, commencing with the producers at the head of the ranking, the threshold was lowered, percentage by percentage, until the amount of quota above the threshold equalled the amount by which the UK had exceeded its quaranteed total quantity. See also, Gehrke, H., *The Implementation of the EC Milk Quota Regulations in British, French and German Law*, at 50–3.

[20] See e.g. *Hansard* (HC) Vol.64, Col.440.　　　[21] [1985] OJ L68/1.

[22] [1984] OJ L90/13.

[23] On such conversion the reserves constituted by Member States for both wholesale and direct sales quota were to be correspondingly increased or reduced.

[24] Case 22/90, *France* v. *Commission*, [1991] ECR I–5285.　　　[25] [1985] OJ L137/5.

The result of the combination of these measures was a significant reduction in the total guaranteed quantity for direct sales quota and corresponding increase in the total guaranteed quantity for wholesale quota. Perhaps the most graphic example was provided by Greece. The initial guaranteed total quantity for direct sales quota amounted to 116,000 tonnes; but by the time of the 1991/2 milk year (allowing for reductions imposed by the Community) the figure was a mere 4,528 tonnes.[26] As noted by the Court of Auditors, conversion on such a scale suggested that allocations of direct sales quota were originally set too high; and the position was exacerbated by delays in fixing the individual reference quantities for direct sales producers.[27]

It may be noted that, in addition to conversion, producers had since the 1984/5 milk year enjoyed the facility of 'quota exchange'.[28] Under the Community legislation the objective of such exchange was to encourage the development of milk production structures by permitting producers who ceased direct sales entirely or in part to receive a wholesale allocation; and producers who ceased wholesale deliveries to receive a direct sales allocation. However, a material constraint was imposed by the need for Member States to have the requisite wholesale or direct sales reference quantities available for allocation. In the United Kingdom it was provided that a producer could exchange direct sales quota for wholesale quota with any other producer in the same region; and this provision survived into the Dairy Produce Quotas Regulations 1991.[29] However, in line with the Community legislation, the efficacy of the measure was restricted by the need to find another producer in the same region who wished, against the trend, to move from wholesale to direct sales production.

1.2.3. Butterfat adjustments

Thirdly, the increase in butterfat levels was causing considerable concern.[30] On the one hand, such an increase was desirable, in that it reflected improvement in the quality of milk. On the other hand, the improvement in

---

[26] See Council Reg. (EEC) 857/84, [1984] OJ L90/13, Annex; and Council Reg. (EEC) 1635/91, [1991] OJ L150/28.

[27] *Special Report No.4/93 on the Implementation of the Quota System intended to control Milk Production*, [1994] OJ C12/1, at 4.22–6 and Annex III.

[28] Comm. Reg. (EEC) 1371/84, [1984] OJ L132/11, Art.4(5) and (6), as originally implemented in the UK by the Dairy Produce Quotas Regs. 1984, S.I.1984 No.1047, reg.8. This facility was not specifically addressed by the Court of Auditors in *Special Report No.4/93*.

[29] S.I.1991 No.2232, reg.16, as authorized by Comm. Reg. (EEC) 1546/88, [1988] OJ L139/12, Art.5(5) and (6). It was not possible to exchange SLOM quota or allocations received under the Nallet Package.

[30] On butterfat issues, generally, see, Milk Marketing Board, *Five Years of Milk Quotas: a Progress Report*, at App. 5; and England and Wales Residuary Milk Marketing Board, *United Kingdom Dairy Facts and Figures: 1994 Edition* (Thames Ditton, Surrey, 1995), at 62.

quality resulted in the same number of litres creating a greater dairy surplus. Indeed, in the period from 1983 to 1990, the average fat content rose from 3.86 per cent to 4.01 per cent of the weight of milk, representing additional production of approximately 178,000 tonnes of butter. In the view of the Court of Auditors (as confirmed by the Commission), to produce 178,000 tonnes of butter would require some 4 million tonnes of milk — total milk production in 1990 being 109 million tonnes.[31] With so much depending on so small a percentage increase, it was perhaps not surprising that, notwithstanding their overtly technical nature, butterfat issues soon became central to the operation of the milk quota system.[32]

The rise in butterfat levels has persisted despite detailed countermeasures. For the purposes of implementing the superlevy Council Regulation (EEC) 857/84 granted the Commission the authority to determine the characteristics of milk 'and, in particular, the fat content thereof, considered to be representative in order to establish the quantities of milk delivered or purchased'.[33] After early difficulties, it was provided that producers and purchasers should receive a fixed butterfat base, in general the average fat content of milk delivered in the 1985/6 milk year.[34] The consequences of exceeding this butterfat base have been significantly amended over time. The legislation in place prior to the reforms was Article 12 of Commission Regulation (EEC) 1546/88,[35] as replaced by Commission Regulation (EEC) 1033/89,[36] the purpose of the change being to simplify the procedure and

---

[31] *Special Report No.4/93 on the Implementation of the Quota System intended to control Milk Production*, [1994] OJ C12/1, at 4.30–4 and Annex XIX. See also the replies of the Commission at 4.34. The Commission, moreover, pointed out that a by-product would be 132,000 tonnes of skimmed-milk powder.

[32] An illustration of this importance is provided by the disputes as to the accuracy of butterfat testing at the close of the 1994/5 milk year (at a time when payment of the superlevy was becoming a virtual certainty): a small upwards discrepancy was capable of transforming an overall national deficit for wholesale quota into a substantial excess — see, e.g., *Farmers Weekly*, 24 Feb. 1995, at 17. The Intervention Board has resisted any allegations as to variation between the different methods employed: *IB Press Notice* 4/95. None the less, since then the monitoring of butterfat testing has become established as a feature of the administration of milk quotas: *IB Press Notice* 13/95.

[33] [1984] OJ L90/13, Art.11(c). See also Comm. Reg. (EEC) 1371/84, [1984] OJ L132/11, Art.9. It may be noted that, accordingly, the butterfat base is only applicable to wholesale quota.

[34] Comm. Reg. (EEC) 1546/88, [1988] OJ L139/12, Art.12; and for first implementation in the UK, the Dairy Produce Quotas Regs. 1989, S.I.1989 No.380, reg.6. It had earlier been provided that the representative characteristics should be those recorded for the previous milk year: Comm. Reg. (EEC) 1371/84, [1984] OJ L132/11, Art.9. In the case of the fixed butterfat base, certain derogations were permitted. Thus, if the fat content of a particular producer or purchaser fell during the 1985/6 milk year, Member States could substitute the average fat content for the 1984/5 milk year. In England and Wales producers whose butterfat content in the 1985/6 milk year was at least 0.10% below the figure for the 1984/5 milk year were granted a butterfat base by reference to the latter period; and this had the effect of increasing the overall butterfat base for the region from 3.96% to 3.98%: Milk Marketing Board, *Five Years of Milk Quotas: a Progress Report*, at App. 5.

[35] [1988] OJ L139/12.   [36] [1989] OJ L110/27.

render it less coercive. These provisions stipulated that, when calculating the superlevy for each producer or purchaser, it was necessary to establish the average fat content of milk delivered for each producer. If this exceeded the producer's butterfat base (as indicated, the butterfat base, in general, being the average fat content of milk delivered in the 1985/6 milk year), then the amount of milk delivered was to be increased by 0.18 per cent for each 0.1 gram of excess fat per kilogram of milk.[37] However, since the enactment of Commission Regulation (EEC) 1033/89, negative as well as positive adjustment had been permitted; and, accordingly, producers whose average fat content was less than their butterfat base could make an equivalent reduction in the amount of milk delivered. To the extent that such reductions took place, latitude was granted to producers who exceeded their butterfat base; and the effect was to introduce further offsetting into the milk quota system, with the responsibility of individual producers for their excess production being again diminished.[38]

1.3. Successful aspects of the milk quota system prior to the reforms

That having been said, evidence may be found that the milk quota system had enjoyed success. Above all a significant fall in production had occurred, notwithstanding the tendency for wholesale deliveries to outstrip national guaranteed total quantities. Thus, the sum of the wholesale guaranteed total quantities for the ten Member States fell from 98,471,600 tonnes in the 1985/6 milk year to 93,313,600 tonnes in the 1990/1 milk year; and, although in each milk year over that period the guaranteed total quantity was exceeded by deliveries adjusted for butterfat content, none the less there was a significant fall from 100,172,900 tonnes to 93,651,300 tonnes.[39] Moreover, at a time when overall European Agricultural Guidance and Guarantee Fund Guarantee expenditure doubled, the level of expenditure in the dairy

[37] On the one hand, the butterfat adjustment had over time become more sensitive. A difference of 0.1 gram per kilogram of milk sufficed to trigger its operation (rather than one gram per kilogram, as had originally been the case). On the other hand, the rate of adjustment had decreased from 0.26 to 0.18% for every 0.1 gram of excess fat per kilogram of milk.

[38] Indeed, following Comm. Reg. (EEC) 1033/89, the milk marketing boards calculated a national butterfat base, being the average butterfat base of all wholesale producers, weighted according to their respective deliveries; and upward adjustment was only triggered when the average butterfat for all milk delivered during the requisite milk year exceeded that national butterfat base: see, e.g., Federation of UK Milk Marketing Boards, *United Kingdom Dairy Facts and Figures:1992 Edition*, at 62. This adjustment has been triggered every year as from the 1989/90 milk year: England and Wales Residuary Milk Marketing Board, *United Kingdom Dairy Facts and Figures: 1994 Edition*, at 62.

[39] See Court of Auditors, *Special Report No.4/93 on the Implementation of the Quota System intended to control Milk Production*, [1994] OJ C12/1, at Annex IV. The fall is rendered the more significant in that the later totals took account of some SLOM allocations and allocations pursuant to the Nallet Package.

sector remained largely static. There had been a sharp rise in 1984, further illustrating the major difficulties then encountered, but thereafter expenditure rose marginally, from 6,295.4 million ECUs in 1984 to 6,786.3 million ECUs in 1991. Indeed, the dairy sector ceased to be the most costly within the Common Agricultural Policy.[40]

With regard to dairy farmers themselves, after initial setbacks their incomes improved considerably as compared with the agricultural sector in general.[41] To attribute this improvement to milk quotas alone would be impossible;[42] but there is no doubt that operators proved — and continue to prove — adept at matching the most profitable production methods to their available quota. For example, there has been a decline in the use of purchased feeding-stuffs, with increased recourse to silage. Further, the number of hired labourers has decreased, and at the same time new technologies have been introduced.[43] Indeed, even as the reforms were being effected, the overall income position of United Kingdom dairy farmers continued to improve. In 1992–3 costs rose by an average of approximately 4 per cent, while milk prices rose by approximately 5 per cent, and calf prices by some 20 per cent.[44]

Alongside this movement in dairy farmers' incomes, there was also clear evidence of structural change (which, as already seen, has been an objective of the milk quota system since its inception, running parallel with the need to curb the dairy surplus).[45] In the United Kingdom automation grew at the expense of hired workers; and, corresponding with the objectives of the

[40] Ibid., at 3.25–8 and Annexes I and XXI. However, the Court of Auditors highlighted that refunds to Member States to dispose of butter in storage were entered elsewhere, which had a potentially distorting effect. For more recent figures, see Residuary Milk Marketing Board, *EC Dairy Facts and Figures 1994* (Thames Ditton, Surrey, 1995), at Table 3.

[41] Eurostat 1990, Vol.1, at 246. For more detailed discussion, see, e.g., Kirke, A. W., 'The Influence of Milk Supply Quotas on Dairy Farm Performance in Northern Ireland', in Burrell, A. M. (ed.), *Milk Quotas in the European Community*, at 30–45; Tollens, E., 'The Effect of Milk Quotas on Community Agriculture, 1984–1987', in ibid., at 183–92; Milk Marketing Board, *Five Years of Milk Quotas: a Progress Report*, at 18–22; and Eurostat 1990, Vol.1, at 77–98 (production and deliveries of milk), 111–36 (profitability of milk production and farmers' incomes), 137–242 (structure of dairy farms and dairies), and 243–8 (summary and conclusion).

[42] For a more detailed analysis of other factors (e.g., a more favourable price ratio of milk to compound feeding-stuffs), see, in particular, Eurostat 1990, Vol.1, at 111–36. See also Harvey, D. R., *Milk Quotas: Freedom or Serfdom?*, at 37–41.

[43] Milk Marketing Board, *Five Years of Milk Quotas: a Progress Report*, at 18–22.

[44] Federation of UK Milk Marketing Boards, *United Kingdom Dairy Facts and Figures: 1993 Edition*, at 9 and 49 and Tables 26–32.

[45] See, e.g., Council Reg. (EEC) 856/84, [1984] OJ L90/10, Preamble. On this aspect, generally, see, e.g., McInerney, J. P. and Hollingham, M. A., *Readjustments in Dairying: an Analysis of Changes in Dairy Farming in England and Wales following the Introduction of Milk Quotas* (Agricultural Economics Unit, University of Exeter, 1989), *passim*; Eurostat 1990, Vol.1, at 137–242 and 247–8; and Court of Auditors, *Special Report No.4/93 on the Implementation of the Quota System intended to control Milk Production*, [1994] OJ C12/1, at 4.58–63.

various outgoers' schemes, there were also significant reductions throughout the Community in the number of dairy farmers themselves, the reductions being most pronounced during the years immediately following the introduction of milk quotas. By way of illustration, the number of dairy farms in the ten Member States fell from 1,621,261 in 1983 to 1,241,627 in 1987 — a reduction of some 23.4 per cent.[46] The United Kingdom has not proved immune from this trend. The number of registered producers dropped from 50,625 in March 1984 to 37,546 in March 1993, while in England and Wales alone the respective figures were 39,287 and 28,729.[47] Although the pace of reduction may now have slackened, it still has not ceased. While this may be seen as no more than the continuation of an existing trend (there having been 60,279 registered producers in England and Wales in 1975), it does suggest that the imposition of the milk quota system did not, in that respect at least, have the effect of ossifying the industry. Outgoers' schemes definitely played a significant role, that contemporaneous with the introduction of milk quotas being directed at small producers; but other factors may be posited. For example, the development of a vigorous market in milk quotas would seem to have hastened the concentration of reference quantities in the hands of larger operators, better placed to raise the requisite capital. At the same time the cost of purchasing reference quantities inevitably created a hurdle for new entrants. That having been said, there is evidence that quotas did impose somewhat of a brake on structural change; and, in any event, there was considerable geographical variation — for example, the rate of decline was far greater in the south-east than the north-west of England.[48]

The number of dairy cows also decreased. For example, in England and Wales the total at the time of the 1984 June census was 2,696,000, while that for the 1992 June census was 2,178,000.[49] However, this decline was partially offset by the improvement in the quality of milk and by an increase in the average annual milk yield per cow. Indeed, in England and Wales the

---

[46] Court of Auditors, *Special Report No.4/93 on the Implementation of the Quota System intended to control Milk Production*, [1994] OJ C12/1, at 4.61.

[47] Federation of UK Milk Marketing Boards, *United Kingdom Dairy Facts and Figures: 1993 Edition*, at Table 1.

[48] See, e.g., McInerney, J. P. and Hollingham, M. A., *Readjustments in Dairy Farming in England and Wales following the Introduction of Milk Quotas*, at 8–14. See also Langer, F. 'Dairy Cessation Schemes, Quota Transfers, and Regional Rigidities', in Burrell, A. M. (ed.), *Milk Quotas in the European Community*, at 149–57 (which highlights the different trends experienced in France, where reference quantities freed by cessation schemes were generally distributed to small or medium-sized producers within a region, so limiting structural change).

[49] Federation of Milk Marketing Boards, *United Kingdom Dairy Facts and Figures:1993 Edition*, at Charts 2 and 3 and Tables 7–10. For details of cow numbers at Community level, see Court of Auditors, *Special Report No.4/93 on the Implementation of the Quota System intended to control Milk Production*, [1994] OJ C12/1, at Annex XVI; and Milk Marketing Board, *EC Dairy Facts and Figures 1993* (Thames Ditton, Surrey, 1993), at Table 13.

average annual yield per cow rose from 4,765 litres in the 1984/5 milk year to 5,175 litres in the 1991/2 milk year.[50] Further, there is evidence that farmers correspondingly increased beef production. In England and Wales the number of beef cattle rose from 701,000 at the time of the 1984 June census to 926,000 by the time of the 1992 June census.[51] Accordingly, it could legitimately be argued that the problems of overproduction were simply transferred to another sector of the Common Agricultural Policy. This in turn may be seen to be a factor behind the major reforms of the beef regimes, also effected in 1992 — and, most notably, the imposition of suckler cow premium quotas.[52]

Finally, there continued to be slow but steady growth in the average size of dairy herds. In England and Wales, for example, the figure grew from 68 in 1986 to 71 in 1992.[53] Moreover, the United Kingdom possessed the largest average herd size in the Community.[54] As a result the overall impression was that of fewer milk producers with fewer cows efficiently producing sufficient milk to equal or marginally exceed their respective reference quantities. In this regard it may be noted that during both the 1991/2 and 1992/3 milk years the level of deliveries to the Milk Marketing Board for England and Wales was such that no superlevy became payable — although, by contrast, it may also be noted that the highest superlevy on wholesale quota to date was subsequently incurred in the 1994/5 milk year.[55]

---

[50] Federation of UK Milk Marketing Boards, *United Kingdom Dairy Facts and Figures: 1993 Edition*, at Table 15. For the equivalent figures at Community level, see Court of Auditors, *Special Report No.4/93 on the Implementation of the Quota System intended to control Milk Production*, [1994] OJ C12/1, at Annex XVII.

[51] Federation of UK Milk Marketing Boards, *United Kingdom Dairy Facts and Figures: 1993 Edition*, at Table 7. See also Eurostat 1990, Vol.1, at 137–242 and 247.

[52] See, generally, Neville, W. and Mordaunt, F., *A Guide to the Reformed Common Agricultural Policy* (Estates Gazette, London, 1993), at 67–102. For the Community legislation, see, in particular, Council Reg. (EEC) 2066/92, [1992] OJ L215/49; and Comm. Reg. (EEC) 3886/92, [1992] OJ L391/20. For the UK legislation, see the Suckler Cow Premium Regs. 1993, S.I.1993 No.1441, as amended by the Suckler Cow Premium (Amendment) Regs. 1994, S.I.1994 No.1528, the Suckler Cow Premium (Amendment) Regs. 1995, S.I.1995 No.15, and the Suckler Cow Premium (Amendment) (No.2) Regs. 1995, S.I.1995 No.1446; the Beef Special Premium Regs. 1993, S.I.1993 No.1734, as amended by the Beef Special Premium (Amendment) Regs. 1994, S.I.1994 No.3131, and the Beef Special Premium (Amendment) Regs. 1995, S.I.1995 No.14; and the Sheep Annual Premium and Suckler Cow Premium Quotas Regs. 1993, S.I.1993 No.1626, as amended by the Sheep Annual Premium and Suckler Cow Premium Quotas (Amendment) Regs. 1993, S.I.1993 No.3036, and the Sheep Annual Premium and Suckler Cow Premium Quotas (Amendment) Regs. 1994, S.I.1994 No.2894.

[53] Federation of UK Milk Marketing Boards, *United Kingdom Dairy Facts and Figures: 1993 Edition*, at Table 14.

[54] See, e.g., Milk Marketing Board, *EC Dairy Facts and Figures 1993*, at Table 14.

[55] For the occasions when the superlevy has become payable in respect of wholesale quota, see England and Wales Residuary Milk Marketing Board, *United Kingdom Dairy Facts and Figures: 1994 Edition*, at 60 (which includes data up to and including the 1993/4 milk year). For

2. THE IMPLEMENTATION OF REFORM

2.1. General

With the passage of time the combination of these factors rendered reform increasingly urgent. Not least, the complexity of the legislation, incorporating a plethora of amendments, strengthened the case for both consolidation and simplification. Such complexity may in part be attributed to Member States furthering national objectives by judicious use of the various discretions contained in the Community legislation.[56] A clear example is provided by the United Kingdom's interpretation of the transfer regulations so as to facilitate a market in quota. However, the detailed nature of the legislation and the wealth of derogations may also be regarded as an inevitable consequence of the twin aims of curbing milk production and, at the same time, permitting structural developments and adjustments.

Agreement that the milk quota system should continue until the year 2000 was reached in the summer of 1992;[57] but detailed reform was not implemented until Council Regulation (EEC) 3950/92 became applicable as from 1 April 1993.[58] The full importance of this new regulation may be judged from the Preamble of Council Regulation (EEC) 2074/92, heralding its enactment, which recited that 'in order to make full use of the experience gained in this area and in the interests of simplification and clarification with a view to ensuring the legal certainty of producers and other parties concerned, the Commission has proposed that the Council lay down by a separate regulation the basic rules of the extended scheme and at the same time reduce their scope and diversity'.[59]

Council Regulation (EEC) 3950/92 confirmed that the milk quota system would remain in place until the year 2000;[60] and in overall framework resembled Council Regulation (EEC) 857/84, which it repealed.[61] However, it also introduced significant modifications. In this context the following may be considered: first, the fixing of national guaranteed total quantities and individual reference quantities (including, in the latter case, the introduction of quota confiscation); secondly, the treatment of quota conversion and butterfat bases (including their role as constituents in the superlevy calculation); thirdly, the superlevy calculation itself; and,

the 1994/5 milk year, see *IB Press Notice* 16/95 (the UK superlevy in respect of wholesale quota being provisionally assessed at £43.4 m).

[56] Court of Auditors, *Special Report No.4/93 on the Implementation of the Quota System intended to control Milk Production*, [1994] OJ C12/1, at 4.10–11.

[57] Council Reg. (EEC) 816/92, [1992] OJ L86/83; and, subsequently, Council Reg. (EEC) 2074/92, [1992] OJ L215/69.

[58] [1992] OJ L405/1.   [59] [1992] OJ L215/69.   [60] [1992] OJ L405/1.

[61] Council Reg. (EEC) 856/84, which inserted the fundamental Art.5c into Council Reg. (EEC) 804/68, remained in force.

fourthly, the overall tightening of superlevy collection and administration procedures.[62]

2.2. Specific measures

2.2.1. National guaranteed total quantities and individual reference quantities

Addressing first the guaranteed total quantities for wholesale quota and direct sales quota, there had been a cut by 2 per cent for the 1991/2 milk year.[63] However, since that date the fixing of definitive guaranteed total quantities had been dependent upon the reform of the Common Agricultural Policy and, by the time that Council Regulation (EEC) 3950/92 was enacted, no final decision had been reached.[64] The necessary adjustment was carried into effect the following year by Council Regulation (EEC) 1560/93.[65] This adjustment largely consolidated the allocations, permanent cuts, and temporary withdrawals to date. In particular, the allocations which had been made from the Community reserve were definitively incorporated into the national guaranteed total quantities (the United Kingdom, accordingly, receiving the 65,000 tonnes allocated in respect of Northern Ireland); and the 4.5 per cent temporary withdrawal was rendered permanent without the payment of compensation.[66] At the same time, as already indicated, the guaranteed total quantity for wholesale quota was increased in the case of certain Member States to permit allocations to, *inter alia*, SLOM producers. The United Kingdom received a 0.6 per cent increase, the wholesale quota guaranteed total quantity for the 1993/4 milk year rising to 14,197,179 tonnes. The guaranteed total quantity for direct sales quota remained at 392,868 tonnes, as for the 1992/3 milk year.[67] These

---

[62] There were also significant reforms to the quota transfer rules, which will be considered in Ch.4.

[63] Council Reg. (EEC) 1630/91, L150/19 (in respect of wholesale quota); and Council Reg. (EEC) 1635/91, [1991] OJ L150/28 (in respect of direct sales quota).

[64] See Council Reg. (EEC) 816/92, [1992] OJ L86/83; Council Reg. (EEC) 3950/92, [1992] OJ L405/1, Preamble; and (for a later interim measure) Council Reg. (EEC) 748/93, [1993] OJ L77/16.

[65] [1993] OJ L154/30.

[66] The temporary withdrawal was first implemented under Council Reg. (EEC) 775/87, [1987] OJ L78/5 (the rate being reduced back to 4.5% by Council Reg. (EEC) 3882/89, [1989] OJ L378/6). It was argued that Council Reg. (EEC) 816/92 and Council Reg. (EEC) 748/93 rendered the temporary withdrawal permanent without compensation and thus gave rise to a claim for non-contractual damages under Art.215(2) of the EC Treaty; but this was rejected by the Court of First Instance: Joined Cases T–466/93, T–469/93, T-473/93, T–474/93, and T–477/93, *O'Dwyer* v. *Council*, (not yet reported). The application for annulment was discontinued.

[67] See also Federation of UK Milk Marketing Boards, *United Kingdom Dairy Facts and Figures: 1993 Edition*, at Table 34. For the subsequent effect of conversion on these totals, see

figures may be compared with respective totals of 15,487,000 tonnes and 398,000 tonnes for the 1984/5 milk year.[68] In the event, some 70,000 tonnes comprised in the 0.6 per cent increase remained unallocated at the close of the 1993/4 milk year; and the United Kingdom sought and obtained authority to divide it among all registered producers on the basis of their reference quantities permanently held as at 31 March 1995. Each received a further 0.5 per cent, which was available to offset any superlevy liability as from the 1994/5 milk year.[69] At the same time as approving this allocation the Agriculture Council rejected the Commission's proposal to cut milk quotas throughout the Community by 1 per cent in the 1994/5 milk year; and confirmed that there would be no reduction in the 1995/6 milk year.[70] Accordingly, producers have been able to look to a period of relative stability in the level of their individual reference quantities.

It may also be mentioned that, as a part of the general reforms, the United Kingdom came to be treated as a single region, with the exception of the Scottish Islands area.[71] Accordingly, there has ceased to be any requirement for regional reserves, the national reserve comprising 'such wholesale and direct sales quota as is not for the time being allocated to any person'.[72]

As regards the fixing of individual reference quantities, Council Regulation (EEC) 3950/92 again effectively consolidated the allocations and reductions to date, providing that such be 'equal to the quantity available on 31 March 1993'. In addition, it was stipulated that these shall be adjusted, where appropriate, so that the sum of the individual reference quantities of the same type do not exceed the corresponding guaranteed total quantities for wholesale or direct sales quota, as the case may be. However, within these guaranteed total quantities, Member States can effect across-the-board reductions to replenish the national reserve for the purposes of

---

Comm. Reg. (EC) 647/94, [1994] OJ L80/16: the guaranteed total quantity for wholesale quota rose to 14,247,283 tonnes, while that for direct sales quota fell to 342,764 tonnes.

[68] Council Reg. (EEC) 1557/84, [1984] OJ L150/6, amending the original figures set by Council Reg. (EEC) 856/84, [1984] OJ L90/10 (in respect of wholesale quota); and Council Reg. (EEC) 857/84, [1984] OJ L90/13 (in respect of direct sales quota).

[69] *MAFF News Releases* 288/94 and 297/94; and *IB Press Notice* 5/95.

[70] *MAFF News Release* 288/94. That having been said, there was an additional cut in the intervention price for butter: Council Reg. (EC) 1881/94, [1994] OJ L197/23.

[71] I.e., any one of (a) the islands of Shetland, (b) the islands of Orkney, (c) the islands of Islay, Jura, Gigha, Arran, Bute, Great Cumbrae and Little Cumbrae, and the Kintyre peninsula south of Tarbert, or (d) the islands in the Outer Hebrides and the Inner Hebrides other than those listed in (c). Quota could not be transferred out of this area: the Dairy Produce Quotas Regs. 1993, S.I.1993 No.923, reg.7(7)(f). See now the Dairy Produce Quotas Regs. 1994, S.I.1994 No.672, regs.7(7), 13(9), and 15(4).

[72] See now the Dairy Produce Quotas Regs. 1994, S.I.1994 No.672, reg.14. For the Community legislation governing the establishment of national reserves, see now Comm. Reg. (EEC) 536/93, [1993] OJ L57/12, Art.6. In particular, the latter requires that there be a separate national reserve for wholesale quota and direct sales quota.

granting additional or specific reference quantities to producers determined in accordance with objective criteria agreed with the Commission.[73]

Council Regulation (EEC) 3950/92 also introduced a new feature of some importance.[74] As from the 1993/4 milk year any producer who does not make deliveries or direct sales or a temporary transfer of quota during a given milk year is liable to suffer confiscation of his reference quantity, which is to be placed in the national reserve. However, if the producer who has suffered confiscation resumes milk production within a time-scale determined by the Member State, he is entitled to the restoration of his reference quantity. In the United Kingdom any such restoration may be made within the period of six years from the commencement of the milk year in which the reference quantity was withdrawn.[75] In the case of wholesale quota purchasers must notify the Intervention Board within forty-five days from the end of each milk year of any producers registered with them who have not made deliveries during that year. It may be noted that, similarly, latitude is not extended to producers under either the Sheep Annual Premium Scheme or the Suckler Cow Premium Scheme. In both those cases continuous leasing is not generally permitted.[76]

2.2.2. *Quota conversion and butterfat bases*

The provisions governing the conversion of quota were relaxed by Article 4(2) of Council Regulation (EEC) 3950/92.[77] Individual producers became entitled to convert any amount of wholesale quota into direct sales quota, and vice versa, on a permanent basis. As a result individual producers are

[73] [1992] OJ L405/1, Arts.4 and 5. Provisional allocations of SLOM 2 quota were specifically addressed in Art.4(3): most notably, it was necessary that they be rendered definitive by 1 July 1993. SLOM 3 quota received no mention, it not being allocated until the enactment of Council Reg. (EEC) 2055/93, [1993] OJ L187/8.

[74] Council Reg. (EEC) 3950/92, [1992] OJ L405/1, Art.5. This provision covers both producers with wholesale quota and producers with direct sales quota.

[75] For the UK implementing legislation, see now the Dairy Produce Quotas Regs. 1994, S.I.1994 No.672, regs.14 and 33, as amended by the Dairy Produce Quotas (Amendment) Regs. 1994, S.I.1994 No.2448, replacing the Dairy Produce Quotas Regs. 1993, S.I.1993 No.923, regs.13 and 32. It is critical that a producer who receives a notification of confiscation from the Intervention Board should within 28 days of receipt notify any person with an interest in the holding; and within 6 months of receipt notify the Intervention Board, *inter alia*, as to whether he wishes to retain the right to restoration: the Dairy Produce Quotas Regs. 1994, S.I.1994 No.672, reg.33(5), as amended by the Dairy Produce Quotas (Amendment) Regs. 1994, S.I.1994 No.2448.

[76] See Comm. Reg. (EEC) 3567/92, [1992] OJ L362/41, Art.7(4) (in respect of sheep quotas); and Comm. Reg. (EEC) 3886/92, [1992] OJ L391/20, Art.34(3) (in respect of suckler cow quotas). In the case of certain environmental programmes the general rule has been relaxed: for the UK legislation, see the Sheep Annual Premium and Suckler Cow Premium Quotas Regs. 1993, S.I.1993 No.1626, reg.10, as amended by the Sheep Annual Premium and Suckler Cow Annual Premium Quotas (Amendment) Regs. 1994, S.I.1994 No.2894.

[77] [1992] OJ L405/1.

no longer restricted to temporary conversion for the currency of a milk year; and it is no longer a prerequisite that the producer hold both wholesale quota and direct sales quota.[78] To facilitate the orderly operation of the system, the producer must submit his application for permanent conversion by 31 December in any milk year, the deadline being delayed in the case of temporary conversion until 28 April in the milk year following the year of temporary conversion. A corresponding increase or reduction is to be made in the wholesale quota of the relevant purchaser. For these purposes the producer must notify the relevant purchaser within seven working days of the conversion; and, where the conversion is permanent, the purchaser whose quota has been increased must notify the Intervention Board within twenty-eight days of the conversion and, in any event, no later than seven working days after the end of the milk year.[79]

The consequences of permitting such widespread conversion have been considerable. A clear illustration is provided by the adjustment in guaranteed total quantities for the 1993/4 milk year effected by Commission Regulation (EC) 647/94.[80] Under that regulation the total quantities definitively converted under Article 4(2) of Council Regulation (EEC) 3950/92 led to an increase in the United Kingdom's guaranteed total quantity for wholesale quota from 14,197,179 tonnes to 14,247,283 tonnes; and a corresponding decrease in the guaranteed total quantity for direct sales quota from 392,868 tonnes to 342,764 tonnes (the latter a material difference in percentage terms).

Three further points may be highlighted. First, notwithstanding the criticism of conversion as reducing the deterrent effect of the superlevy, such provisions have considerably widened its scope. This may be regarded, however, as consistent with the objective of promoting structural change. Secondly, since details of temporary conversion need not be submitted until after the close of the milk year in question, this facility may be seen as a factor liable to cause delays in the calculation of the superlevy. Besides, during the currency of a milk year producers are unable to judge with complete accuracy the national guaranteed total quantities for wholesale and direct sales quota. Finally, while the opportunity for conversion has been extended, that for quota exchange has been removed (its role being effectively overtaken).[81]

---

[78] For the current UK legislation, see the Dairy Produce Quotas Regs. 1994, S.I.1994 No.672, reg.18.

[79] Ibid., reg.6.

[80] [1994] OJ L80/16. In this context it may be noted that the Community legislation provides that purchasers must submit returns in respect of each of their producers by 15 May following the end of the milk year: Comm. Reg. (EEC) 536/93, [1993] OJ L57/12.

[81] The provision governing quota exchange in the Dairy Produce Quotas Regs. 1991, S.I.1991 No.2232 (reg.16) was not re-enacted in the Dairy Produce Quotas Regs. 1993, S.I.1993 No.923.

With regard to butterfat bases, Article 11 of Council Regulation (EEC) 3950/92 provided that detailed rules be adopted as to the characteristics of milk, 'including fat content, which are considered representative for the purposes of establishing the quantities delivered or purchased'.[82] In this it echoes Article 11(c) of Council Regulation (EEC) 857/84, which it replaced.[83] The detailed rules are set out in Article 2 of Commission Regulation (EEC) 536/93.[84] The regulation stipulates that, in general, the characteristics, including the fat content, should be those associated with the individual reference quantities as at 31 March 1993 — as with the level of individual reference quantities, consolidating the position to date. In addition, further provisions address specific situations. Some of these have proved unproblematic, for example the requirement that the representative fat content is not to change where additional reference quantities are allocated from the national reserve. However, others have supplied clear illustration of the difficulties which may be met when seeking to enact simple and clear legislation which at the same time addresses the full range of situations experienced by producers.

Perhaps the most graphic example of the latter category is the treatment of butterfat bases on the conversion of quota. Under Commission Regulation (EEC) 536/93 it was originally provided, *inter alia*, that where conversion gave rise to an *increase* in a wholesale quota reference quantity, then its butterfat base was to remain unchanged. However, where conversion gave rise to the *establishment* of a wholesale quota reference quantity, then its butterfat base was to be the standard figure of 3.8 per cent.[85] The former of these provisions presented an opportunity: a small wholesale quota reference quantity with a high butterfat base (for example, one established by a Channel Island herd) could be increased by large-scale conversion of a direct sales reference quantity; and the increased wholesale quota reference quantity would all enjoy that high butterfat base.[86] To counter such activities Commission Regulation (EC) 470/94 effected an amendment so as to provide that, irrespective whether a wholesale quota reference quantity was increased or established, the converted reference quantity would take on

---

[82] [1992] OJ L405/1.   [83] [1984] OJ L90/13.

[84] [[1993] OJ L57/12. For the detailed Community legislation in force prior to the 1992 reforms, see Comm. Reg. (EEC) 1033/89 [1989] OJ L110/27. Comm. Reg. (EEC) 536/93 also inserted a new proviso: if, following butterfat adjustments, the quantity of milk actually collected in a Member State is greater than the quantity as adjusted, then the superlevy is payable on the difference between the quantity actually collected and the guaranteed total quantity for wholesale quota: Art.2(3). Accordingly, individual negative adjustment cannot result in no payment of superlevy on any quantity collected in excess of the guaranteed total quantity for wholesale quota.

[85] It may be argued that the standard figure of 3.8% worked hard against the UK: e.g., the weighted average butterfat for England and Wales in respect of the 1992/3 milk year was 4.1% — Federation of UK Milk Marketing Boards, *United Kingdom Dairy Facts and Figures: 1993 Edition*, at Table 74.

[86] See, e.g., Comm. Reg. (EC) 470/94, [1994] OJ L59/5, Preamble; and *Farmers' Weekly*, 17 Dec. 1993, at 23.

the standard butterfat base of 3.8 per cent.[87] However, it remains open to producers to show that, exceptionally, the butterfat base of a wholesale quota reference quantity increased by conversion should not be changed.[88]

In addition, Commission Regulation (EEC) 536/93, as now amended by Commission Regulation (EC) 470/94, addresses the effect of quota transfers on the butterfat base;[89] and the butterfat base applicable for producers whose entire reference quantity is derived from the national reserve and who have commenced production after 1 April 1992.[90] As a general rule, the butterfat base for such producers is the average fat content of milk delivered during the first twelve months of production.[91]

However, while these regulations have effected significant changes to the provisions determining an individual producer's butterfat base, the adjustment to be made in the event of him exceeding that base has remained the same: the quantity of milk or milk equivalent is to be increased by 0.18 per cent per 0.1 gram of additional fat per kilogram of milk. Likewise, the equivalent negative adjustment is preserved, should the average fat content over the milk year fall below his butterfat base.[92]

2.2.3. The superlevy calculation

As regards the calculation of the superlevy, the hallmark of the reforms was increased flexibility.[93] However, the high rate applicable to wholesale quota was retained (i.e., 115 per cent of the target price for milk); and, in addition, the rate for direct sales quota was raised to the same figure. This change was justified on the basis that a comparable method of calculating the superlevy was introduced for both wholesale and direct sales quota.[94]

The increased flexibility found clearest expression in the wide scope to 'equal out' the excess production of individual producers over all the reference quantities of the same type in the Member State.[95] It remains the case that the superlevy is payable on wholesale or direct sales production during a milk year in excess of each Member State's respective guaranteed total

---

[87] [1994] OJ L59/5.   [88] Ibid.   [89] For which, see Ch.4, 7.1.2.

[90] Such producers are to be contrasted with those who receive *additional* reference quantities from the national reserve.

[91] Comm. Reg. (EEC) 536/93, [1993] OJ L57/12, Art.2(1)(e), as replaced by Comm. Reg. (EC) 470/94, [1994] OJ L59/5. For implementation of one of the exceptions, see the Dairy Produce Quotas Regs. 1994 S.I.1994 No.672, reg.19.

[92] Comm. Reg. 536/93 [1993] OJ L57/12, Art.2(2).

[93] Strictly speaking, the expression '*superlevy*' has become superfluous following the abolition of the co-responsibility levy by Council Reg. (EEC) 1029/93, [1993] OJ L108/4.

[94] Council Reg. (EEC) 3950/92, [1992] OJ L405/1, Art.1 (and Preamble). In the case of wholesale quota the rate had been raised to 115% by Council Reg. (EEC) 3880/89, [1989] OJ L378/3.

[95] Council Reg. (EEC) 3950/92, [1992] OJ L405/1, Preamble: 'Whereas, in order to keep the management of the scheme sufficiently flexible, provision should be made for individual overruns to be equalled out over all the individual reference quantities of the same type within the territory of a Member State.'

quantities. However, the detailed provisions of Council Regulation (EEC) 3950/92 contain amendments of very considerable importance. First, with regard to wholesale quota, both Formula A and Formula B have been abolished. In their place each Member State, after unused reference quantities have been reallocated or not, can choose to determine the contribution of producers towards the superlevy at the level of the purchaser or at national level. If the former alternative is adopted, the superlevy is calculated 'in the light of the overrun remaining after unused reference quantities have been allocated in proportion to the reference quantities of each producer'. If the latter alternative is adopted, the superlevy is calculated 'in the light of the overrun in the reference quantity of each individual producer'.[96]

The United Kingdom has opted to calculate the superlevy at the level of the purchaser, in a manner which permits offsetting, first, as between producers making deliveries to the same purchaser; and, secondly, as between purchasers. This accords with its earlier adoption of Formula B.[97] In practical terms the calculation has become more complex following the revocation of the milk marketing schemes during the 1994/5 milk year.[98] The five dominant milk marketing boards have now been replaced by a greater variety of purchasers, each of which must carry out the requisite steps to calculate whether they have received excess deliveries. Within this framework the detailed calculation itself is of considerable complexity, not least since it must take into account such aspects as butterfat base adjustments and quota conversion. That having been said, the provisions do ensure the flexibility which is so central to the reforms.

In the case of wholesale quota, the calculation of the superlevy now requires the following steps. First, it is necessary to make any butterfat adjustments to the reference quantities of individual producers. However, as implemented in the United Kingdom, these adjustments are not effected unless the weighted average butterfat content of deliveries made by all producers in the milk year exceeds the base level established for the United Kingdom as a whole. This condition has in fact been satisfied each milk year

---

[96] [1992] OJ L405/1, Art. 2(1).

[97] See now the Dairy Produce Quotas Regs. 1994, S.I.1994 No.672, reg.20 and Sch.5. See also England and Wales Residuary Milk Marketing Board, *United Kingdom Dairy Facts and Figures: 1994 Edition*, at 58–63; and Lennon, A. A. and Mackay, R. E. O. (edd.), *Agricultural Law, Tax and Finance* (Looseleaf, Longmans, London), at F2.4.

[98] Under the Agriculture Act 1993 the Milk Marketing Scheme 1933, the North of Scotland Milk Marketing Scheme 1934, the Aberdeen and District Milk Marketing Scheme 1984, and the Scottish Milk Marketing Scheme 1989 were all revoked on 1 Nov. 1994 (subject to certain specified saving provisions): the Milk Marketing Schemes (Certification of Revocation) (Scotland) Order 1994, S.I.1994 No.2900 (S.146) and the Milk Marketing Scheme (Certification of Revocation) Order 1994, S.I.1994 No.2921. The Milk Marketing Scheme (Northern Ireland) 1989 (reorganized under the Agriculture (Northern Ireland) Order 1993, S.I.1993 No.2665 (N.I.10)) survived until 1 Mar. 1995.

as from 1989/90.[99] Secondly, purchasers are to make temporary reallocations to producers affected by an 'exceptional event' (and if they have
insufficient unutilized reference quantities for this purpose, they must
notify the Intervention Board of the amount of the shortfall). For these
purposes producers affected by exceptional events include producers whose
holdings are in whole or in part placed subject to a notice prohibiting or
regulating the movement of dairy cows pursuant to an order under the
Animal Health Act 1981; and producers whose holdings are situated wholly
or partly within an area which at any time during the milk year has been
designated by an emergency order under section 1 of the Food and Environment Protection Act 1985.[100] Thirdly, the Intervention Board is to determine the amount of excess deliveries made to each purchaser or, as the case
may be, the amount of unutilized reference quantities available to that
purchaser. The determination is made after taking into account quota conversions and the temporary reallocations already mentioned. At the same
time offsetting is achieved between the producers supplying each purchaser, with the unutilized reference quantities allocated to the producers
who have exceeded their reference quantities. Fourthly, any unutilized
reference quantities available to purchasers are added to the national reserve. Fifthly, those reference quantities added to the national reserve are
reallocated, first, to meet any temporary reallocation which has not been
met by a purchaser; and then to any purchasers who have exceeded their
purchaser reference quantities in proportion to their reference quantities.
The basis of this reallocation among purchasers constitutes a material
amendment effected by the Dairy Produce Quotas Regulations 1994.[101]
Prior to that amendment, reallocation among purchasers had been made in
proportion to their excess rather than in proportion to their reference
quantity, which had in effect rewarded over-production. Under the current
provisions, of those purchasers who have exceeded their reference quantity,
it is the purchaser with the greatest reference quantity who receives the
greatest share. Sixthly, the unutilized reference quantities received from the
national reserve are in turn reallocated by the recipient purchasers among
producers who have exceeded their individual reference quantities, again in

---

[99] England and Wales Residuary Milk Marketing Board, *United Kingdom Dairy Facts and Figures: 1994 Edition*, at 62; and *IB Press Notice* 16/95.
[100] For the detailed provisions, see the Dairy Produce Quotas Regs. 1994, S.I.1994 No.672, reg.16. These are highly intricate, covering such aspects as the amount of the temporary reallocation and interaction with quota transfers. It is of note that the authority for temporary reallocation is provided in the context of reallocation of excess superlevy: Council Reg. (EEC) 3950/92, [1992] OJ L405/1, Art.2(4); and Comm. Reg. (EEC) 536/93, [1993] OJ L57/12, Art.5. Indeed, any temporary reallocation under the Dairy Produce Quotas Regs. 1994 is limited to 'an amount of quota corresponding to a proportion of any levy collected in excess of the levy actually due in that year'.
[101] S.I.1994 No.672, Sch.5. See also Consultation Paper issued by MAFF, *Milk Quotas: Proposals to amend the Dairy Produce Quotas Regulations 1993*, at para.27.

proportion to those quantities (rather than the excess). Only at this stage is it possible to make the final calculation of the amount by which each producer has exceeded his individual reference quantity.

As a result it remains a constant that a producer can always be sure of remaining immune from the superlevy, provided that he does not exceed his individual reference quantity. However, even if he exceeds it, several factors absolve him from full responsibility for his over-production — and the reforms have, if anything, increased this latitude. First, if the United Kingdom's guaranteed total quantity for wholesale quota is not exceeded, then the superlevy is not triggered. Secondly, even if the superlevy is triggered at national level, there will be no liability where the total deliveries to the producer's purchaser do not exceed the purchaser's reference quantity.[102] Thirdly, even if the purchaser's reference quantity is initially exceeded, liability may be excluded by reallocations to the purchaser (and, in turn, the producer) received from the national reserve. For these reasons, liability to the superlevy may be seen as potential.[103] However, while the threshold system continues to operate, those producers who are liable must pay at the full 115 per cent of the target price for milk. The scale of this deterrent has been increased by the frequent devaluations of the Green Pound, which have the effect of driving up the rate of the superlevy. Indeed, for the 1994/5 milk year the rate was 30.45 pence per litre — far more than sufficient to make production uneconomic.[104]

The calculation of the superlevy for direct sales is somewhat less complex. There being neither butterfat adjustment nor offsetting at the level of the purchaser, the Intervention Board must first determine the amount of any reference quantities remaining unutilized by individual producers, taking account of conversions between direct sales quota and wholesale quota. Any such reference quantities are then added to the national reserve. Secondly, to the extent reference quantities are available, temporary reallocations are to be made to producers affected by exceptional events.[105] Thirdly, it is necessary to ascertain the amount, if any, by which all direct sales made by individual producers exceed the national guaranteed total quantity, taking into account quota conversions and temporary reallocations. Fourthly, there must be determination of the amount, if any, by which each producer has exceeded his individual reference quantity, again taking into account quota conversions and temporary reallocations. These individual excess sales are then aggregated. Fifthly, the total amount of the superlevy is calculated by

---

[102] With the milk market now open to a greater number of purchasers, it will be of interest to note the extent to which producers are attracted to purchasers who consistently remain within their reference quantities.

[103] See, e.g., England and Wales Residuary Milk Marketing Board, *United Kingdom Dairy Facts and Figures: 1994 Edition*, at 60; and *Faulks* v. *Faulks*, [1992] 15 EG 82, at 85.

[104] *IB Press Notice* 16/95.

[105] As with wholesale quota, these temporary reallocations are made in accordance with the provisions set out in the Dairy Produce Quotas Regs. 1994, S.I.1994 No.672, reg.16.

multiplying any net national excess (as determined at stage three) by 115 per cent of the target price for milk. Finally, it is necessary to calculate the rate of any superlevy to be paid by individual producers on their excess sales (as determined at stage four). This is achieved by dividing the total amount of superlevy by the aggregate of individual excess sales.

Accordingly, in the case of direct sales quota, the superlevy is again a potential liability. As with wholesale quota, if a producer remains within his reference quantity, he will never be liable; and even if he exceeds his reference quantity, the superlevy will not be triggered unless all direct sales exceed the national guaranteed total quantity. Unlike wholesale quota, there is no scope for offsetting at the level of the purchaser; but, by way of compensation, the rate of superlevy may be significantly reduced, since the aggregate of individual excess direct sales is likely to be a considerable amount greater than the net national excess.[106] Indeed, total direct sales did not exceed the national guaranteed total quantity until the 1988/9 milk year; and in that year the rate of superlevy was diluted to a mere 1.4 pence per litre. For the 1992/3 milk year the diluted rate was still just less than 5.6 pence per litre[107] — although on provisional assessment this figure rose dramatically to 16.26 pence per litre for 1994/5.[108]

In this context there is clear advantage in producers being appraised of national production levels over the course of the milk year. In particular, such information provides indication as to whether or not the national guaranteed total quantities for wholesale and/or direct sales quota are likely to be exceeded, so triggering the superlevy. Data provided by the milk marketing boards — and now the Intervention Board — has assisted producers in this respect.[109] However, complete accuracy remains impossible: as seen, the final superlevy liability cannot be determined until after the close of the milk year in question, owing to such factors as quota conversion and temporary reallocation.

[106] In the MAFF Consultation Paper, *Milk Quotas: Proposals to amend the Dairy Produce Quotas Regulations 1993*, there was debate as to whether the dilute levy system should also be applied to wholesale quota. Arguments against such a move (which prevailed) included the following: first, the dilute levy system as originally applied to wholesale quota had been abandoned in favour of the threshold system, on the basis that the threshold system was a more effective method of curbing production; secondly, under the dilute levy system all producers who exceeded their reference quantities faced the same penalty for each litre of excess production, with no punitive element imposed upon producers who exceeded their reference quantities by the greatest amount; and, thirdly, large, efficient producers would be less deterred from exceeding their reference quantities than those operating on a more limited and less profitable scale.

[107] Federation of UK Milk Marketing Boards, *United Kingdom Dairy Facts and Figures: 1993 Edition*, at 60–1.

[108] *IB Press Notice* 16/95. However, this increase may be in part be attributed to the fact that, as from 1 Apr. 1993, the undiluted rate of superlevy for direct sales quota was raised from 75% of the target price for milk to 115% of the target price — in line with the rate for wholesale quota.

[109] See also Court of Auditors' *Special Report No.4/93 on the Implementation of the Quota System intended to control Milk Production*, [1994] OJ C12/1, at 4.19 (which highlights the effective forecasting arrangements adopted by Denmark).

2.2.4. Collection of the superlevy and administration

As indicated, there had been criticism of the general lack of rigour shown by Member States in the collection of the superlevy; this and other administrative matters were firmly addressed in Commission Regulation (EEC) 536/93.[110] Since the 1984/5 milk year notification requirements had been imposed for the purposes of calculating and collecting the superlevy; but the new Commission regulation introduced an altogether stricter framework, reinforced by penalties and verification requirements.[111] By way of draconian example, the reference quantity of a direct sales producer is to be confiscated in the event that he fails to forward before 1 July the requisite declaration summarizing his sales for the previous milk year.[112] Moreover, the influence of the new Community provisions may be seen in the strict deadlines associated with quota transfers towards the end of a milk year.[113]

Finally, over and above these reforms introduced at Community level, significant changes were occasioned in the United Kingdom consequent upon the the revocation of the milk marketing schemes. Whereas the Intervention Board for Agricultural Produce had since the inception of the milk quota system discharged the task of collecting the superlevy to the five milk marketing boards, as from 1 April 1994 this arrangement ceased. Likewise, on the same date it assumed immediate responsibilty for the preparation and maintenance of registers of producers, until then discharged to the boards by the Minister.[114]

---

[110] [1993] OJ L57/12.

[111] The lack of penalties was criticized by the Court of Auditors in *Special Report No.4/93 on the Implementation of the Quota System intended to control Milk Production*, [1994] OJ C12/1, at 4.28. However, prior to Comm. Reg. (EEC) 536/93 the UK already had in place legislation authorizing the payment of interest on overdue superlevy: the Dairy Produce Quotas Regs. 1991, S.I.1991 No.2232, reg.23(3). For verification requirements, see Comm. Reg. (EEC) 536/93, [1993] OJ L57/12, Art.7; and for verification of purchasers prior to the reform of the milk marketing regime, see the Dairy Produce Quotas Regs. 1993, S.I.1993 No.923, regs.6 and 27; and *IB Press Notice* 13/93. In addition, in the UK there have throughout been criminal penalties for certain forms of non-compliance with the national regulations or Community legislation, e.g., for making a statement or using a document when known to be materially false. For the original rules, see the Dairy Produce Quotas Regs. 1984, S.I.1984 No.1047, reg.18; and for the current rules, see the Dairy Produce Quotas Regs. 1994, S.I.1994 No.672, reg.32. In 1995 there were prosecutions for deliveries to unregistered purchasers: see, e.g., *MAFF News Releases* 210 and 243/95; and *IB Press Notice* 18/95.

[112] Reg. (EEC) 536/93 [1993] OJ L57/12, Art.4(2). For the current UK implementing legislation, see the Dairy Produce Quotas Regs. 1994, S.I.1994 No.672, reg.33, as amended by the Dairy Produce Quotas (Amendment) Regs. 1994, S.I.1994 No.2448.

[113] On this aspect, see Ch.4, 7.1.5.

[114] The Dairy Produce Quotas Regs. 1994, S.I.1994 No.672, regs.23 and 25, as amended by the Dairy Produce Quotas (Amendment) Regs. 1994, S.I.1994 No.2448. For reasons of commercial confidentiality it was not thought appropriate that these tasks be entrusted to the milk marketing boards' successor bodies: see, e.g., *MAFF News Release* 196/93. The Intervention Board for Agricultural Produce operates through the Intervention Board Executive Agency, a Government Department under its charge.

4
Quota Transfers

1. INTRODUCTION

1.1. The trade in quota: general

The transfer of milk quotas has proved central to the operation of the quota system. Over and above the practical aspects of such transfers, there is an argument that, on a more theoretical level, they provide a revealing insight into the nature of milk quotas themselves; and, in particular, the relationship of quota with land. In this context it is impossible to ignore the vigorous trade in quota which has developed in some Member States, notably the United Kingdom and the Netherlands. The scale of this trade can be easily be underestimated. For example, the Milk Marketing Board of England and Wales calculated that by the fourth year of the milk quota system (the 1987/8 milk year) some 910,200,000 litres were transferred. For the same year it was estimated that the average price for quota permanently transferred amounted to 28 pence per litre, and that the average price for quota temporarily transferred amounted to 5 pence per litre per annum.[1] More recently, as it became clear that the superlevy would become payable in the United Kingdom for the 1994/5 milk year, prices for both permanently and temporarily transferred quota reached record heights. At the close of the period in which temporary transfers were permitted (i.e., that immediately prior to 31 December 1994), prices of over 20 pence per litre per annum were recorded in the case of temporary transfers and over 70 pence per litre in the case of permanent transfers.[2] Moreover, with the volume of transactions also growing, it has been provisionally calculated that in the 1994/5 milk year some 567 million litres were permanently transferred and some 1,033 million litres temporarily transferred, amounting respectively to 4 per cent and 7.3 per cent of United Kingdom total quota.[3] These prices bore down particularly heavily upon farmers dependent upon temporary transfers for the purposes of commencing or expanding production. Besides,

---

[1] Milk Marketing Board, *Five Years of Milk Quotas: a Progress Report*, at Table 9.

[2] Details of quota prices may be found in the Bruton Knowles Index, compiled at the National Quota Exchange Office in Gloucester, and in trade journals, such as the Markets Section of *Farmers' Weekly*.

[3] *Third Report from the Agriculture Committee: Trading of Milk Quota* (Session 1994–5, H.C.512), at 47.

there was good reason for believing that, for a short period at least, the driving force behind the price of quota was the likely rate of superlevy, as opposed to such factors as long-term estimation of its marginal value in the cost of production or the perceived lifespan of the quota system as a whole.[4]

In the light of these circumstances the motives behind some transfers were called into question. More specifically, there were suggestions that parties unconnected with the dairy industry had been speculating in the value of quota. To address such disquiet the House of Commons Agriculture Committee carried out an investigation into the quota trade, reaching the conclusion that there was no concrete evidence of speculators or quota agents manipulating the market to any significant degree.[5] Rather, it was felt that there had been an unfortunate concatenation of events which led to unusually inflated prices, for example favourable weather in spring leading to unexpectedly high production, expansion to take advantage of the greater sums paid for milk on the revocation of the milk marketing schemes, and, towards the close of the milk year, the near certainty of a significant superlevy.[6] At the same time there was recognition of the flexibility offered by the trade in quotas and, in particular, its role as an engine for structural change.[7] That having been said, there was residual anxiety as to the attendant consequences, even if (as expected) prices fell to more realistic levels. Not least, there was much sympathy for prospective new entrants.[8]

1.2. The trade in quota: the Community view

Such a trade was definitely not an objective of the Community legislature on the introduction of the milk quota system. Instead, transfers were

[4] The economics underlying the value of milk quotas are a subject in themselves: see, e.g., Burrell, A. M., 'The Microeconomics of Quota Transfer', in Burrell, A. M. (ed.), *Milk Quotas in the European Community*, at 100–18; and Hubbard, L. J., *Some Estimates of the Price of Milk Quota in England and Wales* (Department of Agricultural Economics and Food Marketing, University of Newcastle upon Tyne, 1991). For analysis of the position in Ireland, see Conway, A. G., 'The Exchange Value of Milk Quotas in the Republic of Ireland and some Future Issues for EC Quota Allocation', in Burrell, A. M. (ed.), *Milk Quotas in the European Community*, at 119–29.

[5] *Third Report from the Agriculture Committee: Trading of Milk Quota* (Session 1994–5, H.C.512). The Report, together with minutes of evidence and appendices, is a mine of useful information.

[6] Ibid., at xi–xii.

[7] The ability for retiring farmers to turn their reference quantities to account in the market may be regarded as a reason behind the relative lack of importance attached to outgoers' schemes in the UK: see, e.g., Court of Auditors, *Special Report No.4/93 on the Implementation of the Quota System intended to control Milk Production*, [1994] OJ C12/1, at 5.19.

[8] *Third Report from the Agriculture Committee: Trading of Milk Quota* (Session 1994–5, H.C.512), at xv.

perceived as a means of avoiding 'undesirable effects on the structure of milk production', and as a means of providing 'for the possibility of establishing newcomers (such as young farmers)'.[9] Subsequently, in its submissions to the European Court in *Wachauf* v. *Bundesamt für Ernährung und Forstwirtschaft* the Commission openly criticized the development of a free trade in quotas and, in particular, speculation in their value[10] — a view reiterated, for example, in *Ballmann* v. *Hauptzollamt Osnabrück*.[11] This approach also found expression in the regulations already considered to ensure that definitive allocations of SLOM quota were only available to farmers who had the genuine intention and capacity to return to milk production: the allocations were not available to confer 'undue advantage'.[12]

The European Court too was prepared to defend the same principle and for the same reasons. As seen, in the case of *Von Deetzen II* conditions imposed on the transfer of SLOM quota were upheld, in order to prevent the producers concerned receiving a windfall profit.[13] More recently, in *R.* v. *Ministry of Agriculture, Fisheries and Food*, ex parte *Bostock* a similar dislike of speculation in the value of milk quota was used to dispose of the applicant's claim.[14] On the termination of his tenancy he had received no compensation from his landlord in respect of milk quota, the date of termination occurring before any statutory entitlement to such compensation arose under the Agriculture Act 1986.[15] Echoing the judgment in *Von Deetzen II*, the Court stated that 'The right to property safeguarded by the Community legal order does not include the right to dispose, for profit, of an advantage, such as the reference quantities allocated in the context of the common organization of a market, which does not derive from the assets or occupational activitiy of the person concerned'.[16]

It has also been noted that, in the view of the Court of Auditors, the mobility of quotas has lessened the effectiveness of the system as a whole.[17] For example, a producer whose direct sales or deliveries give rise to a genuine risk of incurring the payment of the superlevy may purchase or lease quota to cover any potential excess. A significant factor highlighted in

---

[9] COM(83)500, at 19.   [10] Case 5/88, [1989] ECR 2609, [1991] 1 CMLR 328.
[11] Case 341/89, [1991] ECR I–25.
[12] Council Reg. (EEC) 764/89, [1989] OJ L84/2, Preamble.
[13] Case 44/89, [1991] ECR I–5119, [1994] 2 CMLR 487. It is of note that the Advocate-General observed that there was no need to impose such conditions on the introduction of milk quotas, since milk quotas 'had not yet acquired a market value': ibid., at I–5142 (ECR) and 500 (CMLR).
[14] Case 2/92, [1994] ECR I–955, [1994] 3 CMLR 547.
[15] For greater detail on these aspects, see Ch.5, 4.2.
[16] Case 2/92, [1994] ECR. I–955, [1994] 3 CMLR 547, at I–984 (ECR) and 570 (CMLR).
[17] Court of Auditors, *Special Report No.4/93 on the Implementation of the Quota System intended to control Milk Production*, [1994] OJ C12/1, at 4.35–50.

this context is that the various measures implemented by the national authorities have done much to assist producers in such mitigating strategies. That having been said, the degree of quota mobility varies considerably in the different Member States. As mentioned, the United Kingdom and the Netherlands have operated the most active market;[18] and in the case of the United Kingdom it has been expressly confirmed that the rationale behind such a market is to secure the greatest flexibility within the constraints of the milk quota system and, in consequence, the most economically efficient form of production.[19] At the same time the opportunity for farmers to dispose of reference quantities not required to cover their own production has encouraged maximum use of the national quota. By contrast, in France and Germany a different picture has emerged.[20] Criteria which may have led to their more rigid approach were, in the case of France, the need to ensure a safe passage for the state-controlled restructuring process, with considerable concern that a free market would prejudice young farmers; and, in the case of Germany, the initial over-allocation of reference quantities. In Ireland the operation of 'claw-backs' has acted as a break upon the development of such a market.[21] Thus, where quota was leased after 8 December 1987 legislation authorized deductions from the lessor's reference quantity to the national reserve. The amount of the deduction rose on a graduated scale in line with lessee's reference quantity (with no deduction being applied if the lessee's reference quantity was less than 50,000 gallons). On this basis the legislation may be seen to have encouraged small producers as much as discouraged quoting leasing. It may be noted that no such 'claw-back' or 'siphon' has been implemented in the United Kingdom.[22]

[18] For operation of the market in the Netherlands, see, e.g., De Boer, P. F. W. and Krijger, A., 'The Market for Milk Quotas in the Netherlands with Special Reference to the Correlation between the Price of Land (with Quota) and the Profit per Hectare in Dairy Farming', in Burrell, A. M. (ed.), *Milk Quotas in the European Community*, at 130–48. Cf., the position in Denmark: Walter-Jorgenson, A., 'The Impact of Milk Quotas in Denmark', in ibid., at 21–9.

[19] See, e.g., *Third Report from the Agriculture Committee: Trading in Quota* (Session 1994-5, H.C.512), at xiii (setting out an extract from the evidence of Mr Hollis of MAFF). For such evidence, more fully, see ibid., at 37.

[20] For full and most helpful comparative treatments, see Gehrke, H., *The Implementation of the EC Milk Quota Regulations in British, French and German Law*, at 140–5; and Court of Auditors, *Special Report No.4/93 on the Implementation of the Quota System intended to control Milk Production*, [1994] OJ C12/1, at 4.43–8. See also Lorvellec, L., 'Le régime juridique destransferts de quotas laitiers: commentaire du décret no.87–608 du 31 juillet 1987', [1987] Revue de Droit Rural 409–17.

[21] See, e.g., Geoghegan, H., 'The Superlevy: Sales, Leases and Clawbacks', in *Milk Quotas: Law and Practice: Papers from the ICEL Conference — June 1989*, at 21–6. These clawback provisions were held not unconstitutional in *Condon* v. *Minister for Agriculture and Food*, (1993) 2 IJEL 151. For some time discontinued, clawbacks again form part of the current legislation, the European Communities (Milk Quota) Regs. 1995, S.I.1995 No.266; and see also Notice No.266/2.

[22] For further discussion of this aspect, see Ch.4, 3.

2. BASIC COMMUNITY RULES

2.1. The general principle

From the introduction of the milk quota system, the general principle has been that land transactions are required to effect transfers of quota. In this sense milk quota may be said to be 'linked to the holding'.[23] Such a requirement has been explicitly recognized as furthering the objective already highlighted — the need to prevent a free trade in milk quotas. Indeed, notwithstanding the various amendments and consolidations to the legislation and notwithstanding the various exceptions enacted, the Commission, the Court of Auditors, and the European Court have consistently maintained that the general principle remains intact. For example, in *Wachauf* v. *Bundesant für Ernährung und Forstwirtschaft* the Commission submitted that attachment to the land was a introduced to prevent 'speculative operations'.[24] In like vein, Special Report No.2/87 of the Court of Auditors stated that 'quota can normally only be transferred from one producer to another together with the land to which it relates'; and that 'These restrictions on transfer are intended to prevent a market developing in quota separate from the land.'[25] Moreover, in 1994 the European Court could still speak of the need to ensure that the legislation governing SLOM quota 'enshrines the general principle that every reference quantity is to remain attached to the land in respect of which it is allocated'.[26]

A similar line had already been adopted by national courts. For example, in the Irish case of *Lawlor* v. *Minister for Agriculture* Murphy J accepted the view that quotas were land-based; and even went so far as to say that, in his view, the concept of milk quotas divorced from land would seem meaning-

---

[23] On these aspects, generally, see, e.g., Lawrence, G., 'Milk Superlevy: the Community System', in *Milk Quotas: Law and Practice: Papers from the ICEL Conference — June 1989*, at 1–10; Laffoy, M., 'How are Milk Quotas to be characterised in Irish Land Law? Milk Quotas as Security for Loans', in ibid., at 27–33; Ryan-Purcell, O. *European Community Milk Quota Regulations in Ireland* (unpublished LLM Thesis, University of Limerick, 1992), at Ch.VII; Rodgers, C. P., *Agricultural Law*, at 320–1; Gehrke, H., *The Implementation of the EC Milk Quota Regulations in British, French and German Law*, at 41–2; Cardwell, M. N. and Lane, S., 'The Taxation of Milk Quota', [1994] BTR 501–26; and Snape, J., 'Transfers of Milk Quotas: Law and Tax', (1995) 2 Private Client Business 150–161.

[24] Case 5/88, [1989] ECR 2609, at 2618. See also the submissions of the Commission in Case 44/89, *Von Deetzen II*, [1991] ECR I–5119 (reliance being placed, in particular, on the principle whereby the reference quantity is 'linked to the soil', at I–5134).

[25] [1987] OJ C266/1, at 4.18.

[26] Case 98/91, *Herbrink* v. *Minister van Landbouw, Natuurbeheer en Visserij*, [1994] ECR I–223, [1994] 3 CMLR 645, at 253 (ECR) and 669–70 (CMLR). The Advocate-General could by that date speak of the principle of the 'link with land' as the customary expression, at I–230 (ECR) and 651 (CMLR). See also the Opinion of the Advocate-General in Case 44/89, *Von Deetzen II*, [1991] ECR I–5119, [1994] 2 CMLR 487; and in Case 341/89, *Ballmann* v. *Hauptzollamt Osnabrück*, [1991] ECR I–25.

less.[27] The German Administrative Court was equally forthright in *Re the Küchenhof Farm*, stating that 'The tying of the quota to the land prevents farmers from becoming economically dependent on quota-holders who do not themselves engage in dairy farming and use the provisions of the milk marketing system for commercial purposes which have nothing to do with milk production or the regulation of the market.'[28]

In the United Kingdom the principle of the link with land was endorsed by the High Court decision of *Faulks* v. *Faulks*.[29] The dispute concerned the treatment of milk quota following the death of the younger of two brothers who had farmed in partnership. By the terms of the partnership deed it was stipulated that, on the death of one of partners, a valuation was to be made of the partnership assets and that the surviving partner should enjoy an option to purchase the deceased partner's share. The partnership deed did not specifically address the treatment of milk quota in such circumstances, there being no milk quota system at the time of its execution. However, the deed did specifically address the treatment of the farm tenancy. This, vested in the elder brother in 1983, was to be held on trust for the partners for the duration of the partnership. Accordingly, on dissolution, it was not to be brought into account as an asset of the partnership. The question therefore arose whether the milk quota should pass with the tenancy to the surviving brother or should be valued as a separate partnership asset. The Court held that it passed with the tenancy.

In giving judgment, Chadwick J was prepared to accept that, as a matter of economic reality, milk quota enjoyed an intrinsic value. However, he felt that the general principle whereby milk quota is linked to the holding should apply notwithstanding the existence of derogations from that principle (which will be considered later) and notwithstanding the Inland Revenue's practice of treating milk quota as a separate asset for Capital Gains Tax purposes. Indeed, he expressly rejected the view that the milk quota formed 'an asset separate and distinct from the holding in relation to which it was, or becomes, registered'.[30]

Faulks v. *Faulks* must now be read in the light of *Cottle* v. *Coldicott*, a case heard before the Special Commissioners.[31] In that case milk quota was held to constitute a separate asset for the purposes of the Capital Gains Tax legislation when transferred through the medium of a short-term tenancy.[32] However, the Special Commissioners emphasized the tax context of their decision; and confirmed that for the purposes of the *transfer* rules, the

---

[27] [1990] 1 IR 356, [1988] IRLM 400, [1988] 3 CMLR 22. It may be noted, however, that the judgment was delivered at a time when some of the derogations from the general principle linking milk quota with land were still in their infancy.

[28] [1990] 2 CMLR 289, at 300.   [29] [1992] 15 EG 82.   [30] Ibid., at 88.

[31] [1995] SpC 40.

[32] For transfers through the medium of short-term tenancies, see Ch.4, 3. For the difficulties attendant on such arrangements, see Ch.4, 4.

subject of this chapter, milk quota followed the holding unless one of the authorized derogations applied. Moreover, they stated that on the facts of *Faulks* v. *Faulks* they would have reached the same conclusion and would have held that the quota passed with the tenancy.

2.2. The detailed provisions

As indicated, the Community institutions' objective was carried into effect by a general requirement that milk quota could only be transferred through the medium of a land transaction.[33] Article 7(1) of Council Regulation (EEC) 857/84 as from the very inception of the system stated that 'Where an undertaking is sold, leased or transferred by inheritance, all or part of the corresponding reference quantity shall be transferred to the purchaser, tenant or heir according to procedures to be determined.'[34] Member States were also granted the option to add a part of the reference quantities so transferred to the national reserve.[35] The discretion contained in the original legislation was not exercised in the United Kingdom, in contrast with the position in certain other Member States — notably Ireland, where the 'claw-back' has operated with some severity.

Article 7(1) was supplemented by Article 5 of Commission Regulation (EEC) 1371/84, which laid down the detailed provisions applicable to both wholesale and direct sales quota.[36] On a sale, lease, or transfer by inheritance of an entire holding, the corresponding reference quantity was to be transferred in full to the producer who took over the holding. The rules were more intricate where the sale, lease, or transfer by inheritance affected only part of the holding. In such a case the corresponding reference quantity was to be distributed 'among the producers operating the holding in proportion to the areas used for milk production or according to other objective criteria laid down by Member States'. Member States were, however, authorized to lay down a minimum area necessary to trigger such an apportionment.[37] Somewhat surprisingly, it seemed to be left to the discretion of Member States whether or not to apply these basic transfer

---

[33] The Community rules, however, specified no sanction in the event that the general rule was breached — a point emphasized by the Court of Auditors in *Special Report No.4/93 on the Implementation of the Quota System intended to control Milk Production*, [1994] OJ C12/1, at 4.36.

[34] [1984] OJ L90/13. The expression 'undertaking' was soon replaced by the expression 'holding', consistent with the regulations as a whole: Council Reg. (EEC) 590/85, [1985] OJ L68/1.

[35] Clarification was supplied that wholesale quota would pass to the wholesale quota national reserve and direct sales quota to the direct sales quota national reserve: Council Reg. (EEC) 590/85, [1985] OJ L68/1.

[36] [1984] OJ L132/11.

[37] It was subsequently provided that, should Member States opt to disregard transfers of parts of a holding where the area used for milk production was less than a pre-determined

provisions, the legislation stating that 'Member States may apply' such provisions in respect of transfers taking place during or after the reference period.[38] However, any such discretion did not survive for long;[39] and in any event the United Kingdom enacted regulations in line with the basic provisions.

Finally, the transfer rules were to apply equally to 'other cases of transfer which, under the various national rules, have comparable legal effect as far as producers are concerned'.[40] It was soon established in the case of *Wachauf* v. *Bundesamt für Ernährung und Forstwirtschaft* that the surrender of a lease did have comparable legal effect.[41] As the Advocate-General stated, 'the legal effect of the surrender of a lease must be seen as essentially the same as that of its grant, namely, the transfer of the leased property from one party to the other'.[42]

Special difficulties have arisen in respect of land transactions conducted prior to the introduction of milk quotas on 2 April 1984. The transfer provisions applied equally to transactions which occurred during a Member State's reference year;[43] and the validity of this extended application was challenged in the Irish High Court case of *Lawlor* v. *Minister for Agriculture*.[44] The applicant had sold one of his two dairy farms during 1983, Ireland's reference year. Subsequently, his reference quantity was reduced by 41 per cent, this figure corresponding with the proportion of land used for milk production comprised in the sale. In rejecting the applicant's claim, Murphy J saw it as inherent in the implementing legislation that, should holdings change hands or be subdivided between the reference year and the commencement of the milk quota system, the persons acquiring the holdings should acquire also reference quantities appropriate to the amount and nature of the lands transferred. Indeed, he saw the specific provision contained in the legislation as little more than an '*aide mémoire*' to that effect. However, the European Court adopted a less strict approach in the later case of *Kühn* v. *Landwirtschaftskammer Weser-Ems*, emphasizing rather

minimum, then the part of the reference quantity corresponding to that area might be added in its entirety to the national reserve: Comm. Reg. (EEC) 1546/88, [1988] OJ L139/12, Art.7(2).

[38] Comm. Reg. (EEC) 1371/84, [1984] OJ L132/11, Art.5(3). For judicial consideration, see Case 177/90, *Kühn* v. *Landwirtschaftskammer Weser-Ems*, [1992] ECR I–35, [1992] 2 CMLR 242.

[39] See Comm. Reg. (EEC) 1681/87, [1987] OJ L157/11.

[40] Comm. Reg. (EEC) 1371/84, [1984] OJ L132/11, Art.5(3).

[41] Case 5/88, [1989] ECR 2609, [1991] 1 CMLR 328.

[42] Ibid., at 2627–8 (ECR) and 339 (CMLR). Cf., dicta of McCowan LJ in *W. E. & R. A. Holdcroft* v. *Staffordshire County Council*, [1994] 28 EG 131, at 134: 'I am bound to say that it would not immediately have occurred to me that a surrender of a lease could fall within those words. But the European Court has in fact held, as [Counsel] conceded, that a surrender of a lease does have comparable legal effects.'

[43] Comm. Reg. (EEC) 1371/84, [1984] OJ L132/11, Art.5(3).

[44] [1990] 1 IR 356, [1988] IRLM 400, [1988] 3 CMLR 22.

the discretionary nature of the transfer rules.[45] Thus, it held that where a tenant took over a holding prior to the implementation of the milk quota system, Member States could choose to allocate a reference quantity taking account of deliveries made by the previous tenant during the reference year — but were not obliged to do so.

Accordingly, it is evident that as from 2 April 1984 considerable discretion was granted to Member States in the implementation of the transfer provisions, consistent with the general reluctance of the Community institutions to intervene in the disparate land law régimes of the Member States. More recent illustration of this latitude may be found in the European Court decision of *R. v. Ministry of Agriculture, Fisheries and Food*, ex parte *Bostock*, where it was reaffirmed that the legal relations between landlords and tenants remain governed by national law.[46]

Notwithstanding replacement of the original legislation and the introduction of various derogations, this general requirement for a land transaction to effect a transfer of quota was expressly confirmed when the milk quota regulations were consolidated and clarified in 1992. Indeed, even as it extends the derogations, the Preamble to Council Regulation (EEC) 3950/ 92 in terms refers to 'the principle linking reference quantities to holdings', echoing the expression employed elsewhere by the Community institutions.[47] Further, the Preamble recites that 'when the additional levy system was brought in in 1984, the principle was established that when an undertaking was sold, leased or transferred by inheritance, the corresponding reference quantity was transferred to the purchaser, tenant or heir'; and that 'this original decision should not be changed'. Article 7 (1) of the same regulation re-enacts the principle while at the same time granting Member States wider discretion in its detailed application: 'Reference quantities available on a holding shall be transferred with the holding in the case of sale, lease or transfer by inheritance to the producers taking it over in accordance with detailed rules to be determined by the Member States taking account of the areas used for dairy production or other objective criteria and, where applicable, of any agreement between the parties.' As before, the provisions apply to 'other cases of transfers involving comparable legal effects for producers'. Contained within the same Article is the direction that: 'Any part of the reference quantity which is not transferred with the holding shall be added to the national reserve.'

3. UNITED KINGDOM IMPLEMENTING LEGISLATION

As highlighted, the implementation of this general principle is to be determined in the context of national law. The United Kingdom regulations are

---

[45] [1992] ECR I–35, [1992] 2 CMLR 342.
[46] Case 2/92, [1994] ECR I–955, [1994] 3 CMLR 547. [47] [1992] OJ L405/1.

complex and have given rise to considerable litigation.[48] In addition to securing an appropriate transfer of quota on, for example, the freehold sale or lease of a dairy farm, they encompass sufficient flexibility to permit the operation of the vigorous market in quota.

In carrying into effect Article 7 of Council Regulation (EEC) 857/84 and Article 5 of Commission Regulation (EEC) 1371/84, the initial regulations did not, as noted, provide for the operation of a 'siphon'; and as a result no part of the reference quantity subject to the transfer was added to the national reserve. This remains the case to date, notwithstanding that such a siphon is widely perceived as the potential source of a pool of quota for new entrants — and that under the regimes governing sheep and suckler cow premium quotas the United Kingdom imposes the maximum permitted siphon of 15 per cent on permanent transfers of quota without the holding.[49] However, the House of Commons Agriculture Committee in July 1995 did not yet feel able to recommend the introduction of such a measure, influenced, *inter alia*, by potential prejudice to existing producers.[50]

The initial regulations did, however, lay down in some detail the criteria necessary to trigger a permanent transfer of quota. In particular, it was necessary to demonstrate a 'change of occupation' — an expression not fully defined.[51] For these purposes the United Kingdom exercised its discretion to determine on the transfer of part of a holding the minimum area

---

[48] On transfers in the UK, generally, see, e.g., Wood, D. *et al.*, *Milk Quotas: Law and Practice*, at 17–30; Amies, S. J., 'Transfer of Milk Quota in England and Wales: Present and Preferred Systems', in Burrell, A. M. (ed.), *Milk Quotas in the European Community*, at 158–62; Gregory, M. and Sydenham. A., *Essential Law for Landowners and Farmers* (3rd edn., Blackwell, Oxford, 1990), at 116–20; Rodgers, C. P., *Agricultural Law*, at 312–20; Sir Crispin Agnew of Locknaw Bt, 'Apportionment of Milk Quota', [1992] Journal of the Law Society of Scotland 29–32; Townsend, H., 'The Dangers of Permanently Transferring Milk Quota', (1992) Bulletin of the Agricultural Law Association, Issue 7, 2; Gehrke, H., *The Implementation of the EC Milk Quota Regulations in British, French and German Law*, at 65–74; Federation of UK Milk Marketing Boards, *United Kingdom Dairy Facts and Figures: 1993 Edition*, at 62: Lennon A. A. and Mackay, R. E. O. (edd.), *Agricultural Law, Tax and Finance*, at F2.4; and Moss, J. R., 'Les quotas européens: aspects de l'expérience anglaise et galloise', [1994] Revue de Droit Rural 483–7.

[49] For the Community legislation, see Council Reg. (EEC) 3013/89, [1989] OJ L289/1, Art.5a(4)(b), as inserted by Council Reg. (EEC) 2069/92, [1992] OJ L215/59 (in respect of sheep premium quotas); and Council Reg. (EEC) 805/68, [1968] JO L148/24, Art.4e(1), as amended by Council Reg. (EEC) 2066/92, [1992] OJ L215/49 (in respect of suckler cow premium quotas). For the UK legislation, see the Sheep Annual Premium and Suckler Cow Premium Quotas Regs. 1993, S.I.1993 No.1626, reg.6(1).

[50] *Third Report from the Agriculture Committee: Trading of Milk Quota* (Session 1994–5, H.C.512), at xv and, for the views of MAFF, 35–6. It would seem that, at that date at least, MAFF had doubts, *inter alia*, as to whether a siphon is permitted under Council Reg. (EEC) 3950/92, [1992] OJ L405/1 — presumably as a result of its interpretation of Art.7(1): 'Any part of the reference quantity which is not transferred with the holding shall be added to the national reserve.'

[51] The Dairy Produce Quotas Regs. 1984, S.I.1984 No.1047, Sch.1, Part III (in respect of direct sales quota) and Sch.2, Part III (in respect of wholesale quota).

capable of giving rise to a corresponding transfer of quota. No transfer of quota would arise on a 'minor change of occupation', the definition of which in England and Wales embraced changes of occupation covering an area no larger than five hectares and extending to less than one quarter of the area of the remainder of the holding, provided that the interest of the incoming occupier was a tenancy or less than a tenancy and was for a duration of less than one year.[52] This limitation was dispensed with on the enactment of the Dairy Produce Quotas Regulations 1989.[53]

Following the implementation of the initial regulations, considerable uncertainty remained as to the form of land transfer capable of triggering the requisite change of occupation.[54] This uncertainty was exacerbated by the rapid development of the market in milk quota, for which purposes the parties required the least substantial land transaction consistent with their primary objective of transferring the vendor's reference quantity. Most notably, the practice had already developed of effecting quota transfers through the mechanism of short-term grazing tenancies, or even short-term grazing licences. Under such agreements it was apprehended that a transfer would be triggered on the grant of the grazing tenancy or licence, the reference quantity attached to the land passing to the incoming occupier. The occupier was then required to use the land for non-dairy purposes throughout the term of the agreement (it being usual for the agreement to contain a covenant to graze the land with beef cattle or sheep). On termination it was further apprehended that the land reverted to the grantor without quota, on the basis that it was no longer an 'area used for milk production'.[55] Such arrangements did not fail to attract the attention of the Community institutions; and, in particular, the Court of Auditors in their *Special Report No.4/93* referred to them as a procedure for transferring quota 'without land' — a poignant criticism when the main reason for the arrangements was to secure a transfer *with* land.[56]

---

[52] Ibid., reg.2(1). See also the Dairy Produce Quotas Regs. 1986, S.I.1986, No.470, reg.8; and the Dairy Produce Quotas (Amendment) Regs. 1988, S.I.1988 No.534, reg.5.

[53] S.I.1989 No.380.

[54] E.g., this uncertainty prompted in 1985 the issue by MAFF of a Consultation Paper, *The Mobility of Quota.*

[55] See, in particular, Gregory, M. and Sydenham, A., *Essential Law for Landowners and Farmers*, at 120; Amies S. J., 'Transfer of Milk Quota in England and Wales: Present and Preferred Systems', in Burrell, A. M. (ed.), *Milk Quotas in the European Community*, at 158–62; and Gehrke, H., *The Implementation of the EC Milk Quota Regulations in British, French and German Law*, at 68–70. In the last mentioned of these works such arrangements are referred to as the 'British trick'. See also *R.* v. *Ministry of Agriculture, Fisheries and Food*, ex p. *Cox*, [1993] 22 EG 111, [1993] 2 CMLR 917; and *Harries* v. *Barclays Bank plc* (not yet reported, High Court, 20 Dec. 1995).

[56] [1994] OJ C12/1, at 4.43. See also *Harries* v. *Barclays Bank plc* (not yet reported, High Court, 20 Dec. 1995).

Much of the uncertainty surrounding this trade in quota evaporated on the enactment of the Dairy Produce Quotas (Amendment) Regulations 1988.[57] For the purposes of a transfer of quota by sale, lease, or transfer by inheritance, as a general rule it became necessary that there be a transfer of a holding, or part of a holding, accompanied by a change of occupation. The regulations stipulated certain circumstances in which there would be no transfer of quota. These were, first, a licence to occupy land; secondly, in England and Wales, a tenancy for a period of less than ten months; thirdly, in Scotland, a lease for a period of less than eight months; fourthly, in Northern Ireland, a tenancy for a period of less than twelve months; fifthly, minor changes of occupation; and sixthly, the termination of any such tenancy or lease. As already mentioned, the fifth of these categories (minor changes of occupation) was not carried through into the Dairy Produce Quotas Regulations 1989; but the remainder are still applicable by virtue of Regulation 7(6) of the Dairy Produce Quotas Regulations 1994.[58] It has astutely been pointed out that under the Community regulations there is no authority for distinguishing between tenancies of very short-term duration which do not effect a transfer of quota and tenancies of greater length which do have that effect.[59]

While the Dairy Produce Quotas (Amendment) Regulations 1988 did little to limit the vigour of the market in milk quota, they did produce a higher degree of certainty as to the effect on quota of differing forms of land transaction. Not least, by the grant of a tenancy for a term of less than ten months landowners could let land used for milk production without fear of claim on their reference quantity at the termination of the tenancy.

Accordingly, since the implementation of the Dairy Produce Quotas (Amendment) Regulations 1988 a transfer of quota has arisen on freehold sales, on transfers by inheritance, and on the grant and termination of certain tenancies. Under the AHA 1986, applicable in general to tenancies granted prior to 1 September 1995,[60] the following forms of tenancy triggered such a transfer: first, tenancies which attracted the full security of the

---

[57] S.I.1988 No.534, reg.5 (prior to which there had been further consultation). For a clear expression of Ministerial concern, see, e.g., *MAFF News Release* 351/87.

[58] S.I.1994 No.672. In addition, the 1994 Regs. clarified the correct treatment of quota on the termination of a licence. Under the earlier regulations it had not been stated in terms that there would be no transfer of quota; but the 1994 Regs. expressly provide that there will be no transfer in such circumstances: the Dairy Produce Quotas Regs. 1994, S.I.1994 No.672, reg.7(6)(b).

[59] Hill and Redman's, *Law of Landlord and Tenant* (Looseleaf edn., Butterworths, London), at Para.F[5221]. See also the expert advice of Proferror Usher referred to in the County Court case of *Carson* v. *Cornwall County Council*, [1993] 03 EG 119 (which questioned whether an agreement for less than a milk year could be effective to transfer quota).

[60] For the exceptional circumstances in which the AHA 1986 still applies to tenancies granted on or after 1 Sept. 1995, see the ATA 1995, s.4.

AHA 1986;[61] secondly, fixed-term tenancies for a period of two years or more which continued as fully protected tenancies under section 3 of the AHA 1986; thirdly, fixed-term tenancies of more than one but less than two years (i.e., *Gladstone* v. *Bower* agreements[62]); and, fourthly, Ministry-consent tenancies granted under section 5 of the AHA 1986 (i.e., for a fixed term of not less than two but not more than five years). In addition, transfers were triggered by Ministry-consent tenancies, but not licences,[63] granted under section 2 of the AHA 1986 (i.e., for a fixed term of less than a year); and also by grazing and/or mowing tenancies, but not licences, for a specified period of the year. However, in both these cases it was a requirement that the term of the agreement be for a period of at least ten months.[64] It may be noted that *Gladstone* v. *Bower* agreements and grazing and/or mowing tenancies found particular favour in this context. As with Ministry-consent tenancies, they conferred no security of tenure. However, unlike Ministry-consent tenancies, there was no need to apply to the Minister, whose consent in any event was unlikely to be forthcoming: the transfer of milk quota did not fall within the criteria justifying approval set out in the joint announcement by the Agriculture Departments for England and Wales of 10 August 1989. None the less, where the primary objective was the transfer of quota, there is no doubt that under the AHA 1986 very considerable care was needed to ensure that, on the one hand, the land transaction proved effective to achieve that objective; and that, on the other hand, it did not grant security of tenure.

The position has been somewhat simplified following the enactment of the Agricultural Tenancies Act 1995 and the consequent introduction of the farm business tenancy régime as from 1 September 1995. The land transaction must still be effective to trigger a transfer of quota; but the grantor can enter into the requisite tenancy for a period of ten months of more safe in the knowledge that the grantee will receive only the reduced level of security confered by the new legislation.[65]

---

[61] This included any licence converted into a fully protected tenancy by virtue of the AHA 1986, s.2.

[62] [1960] 2 QB 384 [1960] 3 All ER 353, [1960] 3 WLR 575.

[63] It was questionable in any event to what extent such a licence could be granted at all: see Densham, H. A. C., *Scammell and Densham's Law of Agricultural Holdings*, at 49; and *Ashdale Land & Property Co. Ltd.* v. *Manners*, [1992] 34 EG 76.

[64] More detailed consideration of the landlord and tenant aspects may be found in, e.g., Muir Watt, J., *Agricultural Holdings*, at 9–27; Densham, H. A. C., *Scammell and Densham's Law of Agricultural Holdings*, at 30–3, 35–41, and 44–9; and Rodgers, C. P., *Agricultural Law*, at 25–38 and 316–7.

[65] On the farm business tenancy régime, generally, see Evans, D., *The Agricultural Tenancies Act 1995* (Sweet and Maxwell, London, 1995); Moody, J. with Jessel, C., *Agricultural Tenancies Act 1995 — a Practical Guide* (Farrer and Co., London, 1995); and Sydenham, A. and Mainwaring, N., *Farm Business Tenancies: Agricultural Tenancies Act 1995* (Jordans, Bristol, 1995).

4. DIFFICULTIES WHICH HAVE ARISEN ON QUOTA TRANSFERS

4.1. General

Notwithstanding the clarification provided by the Dairy Produce Quotas (Amendment) Regulations 1988, the transfer provisions on sale, lease, or transfer by inheritance have not proved free from difficulty. This has been especially so in the case of transfers through the medium of short-term agreements. For example, as indicated, the Court of Auditors view these as failing to satisfy the requirement for a land transaction. Moreover, it has also been seen that in *Carson* v. *Cornwall County Council* the County Court accepted Professor Usher's expert advice that the land comprised in such a short-term agreement (which the occupier is contractually bound not to use for milk production) should be characterized as falling outside the definition of 'holding' altogether — rather than as land which constitutes part of a holding, but a part not qualifying as an area used for milk production.[66]

4.2. Specific difficulties

In this context four specific difficulties may be addressed. First, although the regulations require that there be a transfer of all or part of a holding accompanied by a change of occupation, uncertainties have arisen consequent upon the absence of a full definition of the expression 'change of occupation'.[67] In particular, where the primary objective of the transaction is to secure the transfer of quota, there may be little incentive for the transferee to take physical occupation of the land, and even less incentive if the transferor's land is far distant from his own. The position is complicated where, to overcome these practical considerations, the transferee appoints the transferor to act as his contractor during the currency of the agreement.[68]

---

[66] [1993] 03 EG 119. For similar expert doubt as to the efficacy of these short term agreements, see Densham, H. A .C., *Scammell and Densham's Law of Agricultural Holdings, 1993 Supplement*, at A20–1. See also *Faulks* v. *Faulks*, [1992] 15 EG 82 (where Chadwick J questioned the extent to which it was possible to acquire quota 'on the back of a transfer of a limited interest in land': at 95). In *Cottle* v. *Coldicott* the Special Commissioners felt that they must proceed on the basis such transactions are effective; but a certain scepticism may be detected: [1995] SpC 40. Cf., the approach adopted to the trade in banana import licences: Case 280/93, *Germany* v. *Council*, [1994] ECR I–4973.

[67] For the current legislation, see the Dairy Produce Quotas Regs. 1994, S.I.1994 No.672, reg.7. This necessity for a change of occupation is reinforced by the definition of 'transferee', in circumstances where quota is transferred with land, as 'a person who replaces another person as occupier of a holding or part of a holding'. Likewise, in such circumstances a 'transferor' is defined as 'a person who is replaced by another person as occupier of a holding or part of a holding': ibid., reg.2(1).

[68] For evidence of prevailing uncertainty, see, e.g., the Ministry guidelines issued before the Dairy Produce Quotas (Amendment) Regs. 1988, came into force. These stated, *inter alia*, that:

The potential dangers were thrown into sharp relief by the decision of the High Court in *R. v. Ministry of Agriculture, Fisheries and Food*, ex parte *Cox*.[69] On the facts there had been an agreement to transfer 65,000 litres of milk quota through the medium of an eleven-month grazing licence.[70] The transferee at no point took physical possession of the land. Moreover, the transferor took possession during the fifth month of the agreement. Following termination of the licence, with the reference quantity now registered in the name of the transferee, the transferor claimed that the transfer was invalid. After close consideration of the Community and United Kingdom regulations, the Court decided that the absence of physical occupation did indeed prevent a valid transfer.[71] In particular, Article 5(1) of Commission Regulation (EEC) 1371/84 required that, where the entire holding was sold, leased, or transferred by inheritance, the corresponding reference quantity passed to the producer *who took over the holding*.[72] Further, under Article 5(2) of Commission Regulation 1371/84, where part of the holding was sold, leased, or transferred by inheritance, the corresponding reference quantity was to be distributed among the producers *operating the holding*. With regard to the United Kingdom regulations, Popplewell J referred to the definition of 'occupier', which had been extended to include, where there was no occupier, the person entitled to grant occupation; and such an extension was felt unnecessary unless 'occupier' meant 'somebody in physical occupation'.[73] Accordingly, he rejected the argument that the right to occupy was sufficient: what was required was 'the physical use of the land to some degree'.[74] The full import of this test is not yet clear — although some guidance was provided in the course of the judgement. Popplewell J en-

'The Community Regulations themselves are not completely clear on the extent to which a transferee needs to exercise his rights over the land transferred to him, but the Government takes the view that the transferee should have at least an exclusive right of occupation. Where no physical occupation takes place, it is likely that a full explanation of the circumstances will need to be provided to the Ministry': at para.9.

[69] [1993] 22 EG 111, [1993] 2 CMLR 917.

[70] The agreement was made on 10 Sept. 1986, before the Dairy Produce Quotas (Amendment) Regs. 1988 precluded the transfer of quota through the medium of grazing licences.

[71] The Court also decided that the transferee should none the less retain the reference quantity, since, although the Minister had the power to alter the registration, as a matter of administrative law it was felt inequitable that he should do so. A factor leading to this decision was the presence of the Ministry guidelines issued before the Dairy Produce Quotas (Amendment) Regs. 1988 came into force (on which, see above). The parties had conducted the transaction according to common practice and on the basis that the steps taken were sufficient as a matter of law; and, in the light of the guidelines, the Court felt this not surprising. At the same time there was a reluctance to see the transferor deprived of her reference quantity and, consequently, liable to heavy superlevy payments.

[72] [1984] OJ L132/11.

[73] [1993] 22 EG 111, [1993] 2 CMLR 917, at 114 (EG) and 926 (CMLR). The definition applicable was that in the Dairy Produce Quotas Regs. 1986 S.I.1986, No.470, reg.2(1). It remains unaltered in the Dairy Produce Quotas Regs. 1994, S.I.1994 No.672, reg.2(1).

[74] [1993] 22 EG 111, [1993] 2 CMLR 917, at 114 (EG) and 926 (CMLR).

quired whether putting one sheep on the land for the day would be sufficient — a proposition which was eventually rejected by counsel, who contended that more than *de minimis* use was necessary.

In these circumstances analogy could be drawn with the nature of the
occupation required for the purposes of the Occupiers' Liability Act 1957
and for the purposes of the Landlord and Tenant Act 1954. The former
statute emphasizes the importance of control. Exclusive occupation is not a
necessity, nor is it a requirement that the occupier have an estate in land.
Moreover, the occupier may be liable even where he has, by contract,
allowed a third party to have use of the premises.[75] Similarly, in the case of
the Landlord and Tenant Act 1954 control is again of central importance.[76]
That having been said, in *Graysim Holdings Ltd.* v. *P. & O. Property
Holdings Ltd.* the House of Lords recently affirmed that the word 'occupied' carried the connotation of some physical use of the property by the
tenant for the purposes of his business.[77]

While the House of Lords in *Graysim* also cautioned that the meaning of
such expressions as 'occupied' and 'occupation' depend on their context,
the judgment is not likely to help, for example, the transferee who wishes to
place a contractor on the tenanted land. Further, the early milk quota
legislation made express reference to producers 'operating' the holding,
which would seem to imply some degree of activity;[78] and even the more
neutral reference to producers 'taking over' the holding, as found in the
current regulations, could be said to import more than control.[79]

In addition, failure to take physical occupation could create a risk that the
transaction be considered a 'sham' or 'pretence' at common law, without
reference to the Community or United Kingdom regulations.[80] This possibility was also entertained in *R.* v. *Ministry of Agriculture, Fisheries and
Food*, ex parte *Cox*, the Court expressly noting that the Milk Marketing
Board had the power to prevent sham transfers. However, it was stated that
'if grazing by sheep for a sufficiently substantial period will enable a milk

---

[75] See, e.g., *Wheat* v. *Lacon and Co. Ltd.*, [1966] AC 552; and, generally, Rogers, W. H. V.,
Winfield and Jolowicz's Law of Torts (14th edn., Sweet and Maxwell, London, 1994), at 227–8.
[76] See, e.g., *Groveside Properties Ltd.* v. *Westminster Medical School*, (1983) 47 P & CR 507.
[77] [1995] 3 WLR 854.
[78] Comm. Reg. (EEC) 1371/84, [1984] OJ L132/11, Art.5(2).
[79] Council Reg. (EEC) 3950/92, [1992] OJ L405/1, Art.7.
[80] See, in particular, Court of Auditors, *Special Report No.2/87 on the Quota/Additional
Levy System in the Milk Sector*, [1987] OJ C266/1, at 4.19: 'the Court noted that controls on
whether or not authorised land deals with quota were genuine (for example, whether or not a
lessee actually occupied the land which he had leased with quota) were either limited in extent
or non-existent'; and Rodgers, C. P., *Agricultural Law*, at 317–8. For a general statement as to
the nature of sham transactions, see *Snook* v. *London & West Riding Investments Ltd.*, [1967]
2 QB 786; and, for a more recent view in the context of residential lettings, see *Kaye* v.
Massbetter Ltd., [1991] 39 EG 129. See also Bright, S., 'Beyond Sham and into Pretence' (1991)
11 OJLS 136–45.

quota to be transferred that agreement could scarcely be described as a sham'.[81] None the less, there would seem to be scope for alleging a 'sham' or 'pretence' in circumstances where the degree of occupation falls below the requisite level; and in this context the risk may be magnified where the transferee appoints the transferor as his contractor during the currency of the agreement.[82]

Secondly, there is an argument that, at least under the early legislation, use for non-dairy purposes throughout the limited period of a grazing and/ or mowing agreement, or even a *Gladstone* v. *Bower* agreement, would be insufficient to ensure the transfer of all of the quota. The question would become particularly relevant where part only of the holding was transferred and a dispute regarding the apportionment of the quota proceeded to arbitration. Under the early legislation[83] the arbitrator was required in these circumstances to base his award on findings made by him as to areas used for milk production in the *five* years preceding the change of occupation. If, therefore, a short-term agreement had precluded dairy use for only eleven months, it was questionable whether there had been a transfer of all of the quota attaching to the land as at the grant of the agreement (as opposed to a proportion of that quota calculated upon a time basis).[84] However, the effect of these provisions has been mitigated by amendment pursuant to the Dairy Produce Quotas Regulations 1993.[85] Since then the arbitrator has been directed to base his award on findings made by him as to areas used for milk production in the last five year period during which production *took place*.[86]

Thirdly, although the Dairy Produce Quotas (Amendment) Regulations 1988 made clear that the grant of a licence is ineffective to trigger a transfer of quota, the operation of this provision in national law has not always been free from uncertainty. Not least, there are good grounds for holding that a

---

[81] [1993] 22 EG 111, [1993] 2 CMLR 917, at 114 (EG) and 925 (CMLR).

[82] In addition, under the AHA 1986 grazing and/or mowing agreements were only exempt from security of tenure when made in contemplation of the use of the land for grazing or mowing (or both) during some specified period of the year: s.2(3)(a). It has been pointed out that many agreements employed for the purposes of transferring quota may well have lacked such contemplation, with the result that the transferor could inadvertently have granted a protected tenancy: Rodgers, C. P., *Agricultural Law*, at 317–8.

[83] I.e., all legislation up to and including (in the case of England and Wales) the Dairy Produce Quotas Regs. 1991, S.I.1991 No.2232, Sch.4, para.3.

[84] It could also be argued that the land comprised in the agreement had, on the termination of the agreement, ceased to be part of the Euro-holding altogether: see the expert of advice of Professor Usher accepted by the County Court in *Carson* v. *Cornwall County Council*, [1993] 03 EG 119.

[85] S.I.1993 No.923. For the current provisions in this form (in the case of England and Wales), see the Dairy Produce Quotas Regs. 1994, S.I.1994 No.672, Sch.2, para.3(1).

[86] See also *Carson* v. *Cornwall County Council*, where the County Court accepted Professor Usher's doubts as to whether an agreement for less than a milk year would be effective to transfer quota: [1993] 03 EG 119.

licence has limited scope in such a context, since, where it confers exclusive possession for a term at a rent, a tenancy will generally arise in accordance with the principle enunciated in *Street* v. *Mountford*.[87] Further, in the case of agricultural tenancies granted before 1 September 1995 there is the additional factor that such a licence has generally been converted into a fully protected tenancy under section 2 of the AHA 1986.[88] That having been said, it is strongly arguable that no transfer of quota would occur where, for example, a landowner grants himself and another a non-exclusive licence over his freehold property for the purposes of a partnership business.[89] Other examples might be a non-exclusive licence granted by a landowner entering into a share-farming agreement;[90] or a gratuitous licence.[91]

Fourthly, problems confront persons not actively engaged in milk production who seek to purchase quota through the medium of short-term tenancies. As indicated, it is the essence of such transactions that production should not occur on the land comprised in the tenancy. However, production *should* occur on other land in the transferee's Euro-holding, so that on termination of the tenancy agreement the quota is apportioned to that other land. If there is *no* production, whether on the land comprised in the tenancy or on any other land in the Euro-holding, such a result would seem hard to sustain.

4.3. Business reorganizations

Over and above these specific difficulties, more general doubts exist as to the correct treatment of milk quota where business structures undergo change. In the United Kingdom decision of *Faulks* v. *Faulks* it has already been seen that, on the death of one farming partner, the quota passed with the tenancy to the survivor, since under the partnership deed the deceased's estate enjoyed no interest in the tenancy.[92] The Court seemed to treat the

---

[87] [1985] AC 809, [1985] 2 All ER 289, [1985] 2 WLR 877.

[88] For judicial consideration of the interaction between *Street* v. *Mountford* and the AHA 1986, s.2, see *McCarthy* v. *Bence*, [1990] 17 EG 78. See also Densham, H. A. C., *Scammell and Densham's Law of Agricultural Holdings*, at 51; Rodgers, C. P., *Agricultural Law*, at 31–2; and Rodgers, C. P., [1991] Conv. 58–65.

[89] See, e.g., *Harrison-Broadley* v. *Smith*, [1964] 1 All ER 867, [1964] 1 WLR 456; and *Bahamas International Trust Co. Ltd.* v. *Threadgold*, [1974] 3 All ER 881, [1974] 1 WLR 1514. See also, e.g., Slatter, M. and Barr, W., *Farm Tenancies*, at 28–30; Muir Watt, J., *Agricultural Holdings*, at 11 and 17–18; Densham, H. A. C., *Scammell and Densham's Law of Agricultural Holdings*, at 51–2 and 59–60; and Rodgers, C. P., *Agricultural Law*, at 41–2.

[90] See, e.g., *McCarthy* v. *Bence*, [1990] 17 EG 78. See also, e.g., Densham, H. A. C., *Scammell and Densham's Law of Agricultural Holdings*, at 53–4; and Rodgers, C. P., *Agricultural Law*, at 39–41.

[91] See, e.g., *Goldsack* v. *Shore*, [1950] 1 KB 708, [1950] 1 All ER 276. See also, e.g., Muir Watt, J., *Agricultural Holdings*, at 11; Densham, H. A. C., *Scammell and Densham's Law of Agricultural Holdings*, at 50–1; and Rodgers, C. P., *Agricultural Law*, at 27–8 and 38.

[92] [1992] 15 EG 82.

determination of the beneficial interest under the trust as a case of transfer which, under the disparate national rules, had comparable legal effect for producers as a transfer by inheritance.[93]

Similar questions had earlier been addressed by the European Court in the SLOM quota case of *Von Deetzen II*.[94] In addition to addressing the validity of the transfer restrictions imposed on SLOM quota, the European Court was required to decide the effect of those restrictions on the applicant's plans to retire from farming. More specifically, he did not wish those plans to give rise to a 'sale' or 'lease' such as would cause the quota to be returned to the Community reserve.[95] Instead, it was the applicant's aim to show that the arrangement amounted to no more than the transfer of the holding 'by inheritance or by any similar transaction', in which case the SLOM quota would follow the land under the general rules (provided the producer taking over the holding undertook to comply with his predecessor's undertakings).[96] The Advocate-General emphasized that, in the case of such hypothetical questions, where the precise details were obscure, the European Court could 'give only general guidelines on the approach to be adopted by the national court'.[97] However, he also emphasized the distinction between transactions with the sole objective of continuing the farming operation and transactions enabling the quota holder to realize its value. The latter, he felt, would lead to forfeiture of the quota, whether there be sales, leases, or, indeed, transactions with comparable legal effect. Addressing this question with regard to the formation of a partnership, the Advocate-General reaffirmed that much would depend on the national court's interpretation of the relevant partnership agreement, the criterion being 'whether the formation of the partnership will lead to the exchange of quota, or of the profits derived from working the quota, for a share in other profits or assets'.[98] If this criterion was satisfied, the transaction would effectively result in the realization of the commercial value of the quota and give rise to forfeiture to the Community reserve. Moreover, in his view, the criterion would generally be satisfied in the case of commercial arrangements between partners dealing at arm's length. By contrast, in the case of a partnership with prospective heirs, the transaction could qualify as 'similar to inheritance', the Advocate-General expressly accepting that transactions similar to inheritance could include lifetime rearrangements. Equally,

---

[93] Ibid., at 85, referring, in particular, to Comm. Reg. (EEC) 1371/84, [1984] OJ L132/11, Art.5(3).

[94] Case 44/89, [1991] ECR I–5119, [1994] 2 CMLR 487.

[95] Council Reg. (EEC) 857/84, [1984] OJ L90/13, Art.3a(4), as inserted by Council Reg. (EEC) 764/89, [1989] OJ L84/2.

[96] Comm. Reg. (EEC) 1546/88, [1988] OJ L139/12, Art.7a, as inserted by Comm. Reg. (EEC) 1033/89, [1989] OJ L110/27.

[97] Case 44/89, [1991] ECR I–5119, [1994] 2 CMLR 487, at I–5143 (ECR) and 501 (CMLR).

[98] Ibid., at I–5144 (ECR) and 503 (CMLR).

while most transmissions of interests on death would not cause the SLOM
quota to revert to the Community reserve, this general rule would not apply
if the transmission was an incident of a purely commercial arrangement —
in which case forfeiture would again occur. In this context the focus of the
Opinion was directed to the accrual of a deceased partner's share to surviving partners; and it was emphasized that there could easily be circumstances
where this amounted to a 'sale', whether at the date of death or earlier on
the formation of the partnership.

The European Court essentially followed this approach. A 'sale' or
'lease' was to be construed 'as referring to any transfer, for consideration, of
the ownership of or right to use the holding, whatever the legal basis of the
transfer', with the exception of cases where it was the result of a transaction
similar to inheritance. Such a 'similar transaction' referred to any transaction 'whatever its legal basis, which produces effects comparable to those of
inheritance', for instance transactions between a producer and his potential
beneficiary, so long as the primary purpose was to continue farming rather
than to realize the marketable value of the holding.[99] While this decision
addressed the circumstances in which SLOM quota might be forfeited, it is
arguably of more general interest. Most notably, the Court looked to the
effect of the transaction in commercial terms; and was prepared to look
beyond any narrow definition of 'sale', 'lease' or, indeed, transaction 'similar to inheritance'. In this light there are grounds for believing that the
Court would consider the purpose of any business rearrangement, rather
than the precise form of documentation employed; and, if the purpose is to
permit a member of the business to realize the value of the quota, then a
sale or lease may be deemed to have occurred.

Subsequent SLOM litigation emphasized these problems. In particular,
partnerships applying for SLOM allocations were often differently constituted from those which had entered into the non-marketing or conversion
undertakings, owing to the considerable lapse of time which had inter-
vened. When faced by such circumstances the European Court has adopted
a pragmatic approach. As already indicated, in *R.* v. *Ministry of Agriculture,
Fisheries and Food*, ex parte *Dent* it was held that, where a producer
had received an award of exceptional hardship quota, this fell to be deducted from a later SLOM allocation under the 'anti-cumulation rule'.[100]
However, in this context it was further held that the anti-cumulation
rule would still apply in circumstances where the exceptional hardship
award had been made to three individuals as a partnership, but the SLOM
allocation was made in favour of just the two partners who had originally
entered into the conversion scheme. The Court would seem to have looked

[99] Ibid., at I–5159 (ECR) and 512 (CMLR).
[100] Case 84/90, [1992] ECR I–2029, [1992] 2 CMLR 597.

through the change in business structure with a view to preventing a double allocation of quota — emphasizing that the allocations should have been made to the 'partnership as such'.[101] Likewise, in *O'Brien* v. *Ireland* the Advocate-General was prepared to see a SLOM allocation made to a partnership notwithstanding that the farmer who had entered into the non-marketing undertaking had traded on his own account.[102] The change in business structure was not felt to be an insuperable barrier, although it was necessary that the original sole trader be a member of the partnership.[103]

More recently, in the United Kingdom case of *W. E. & R. A. Holdcroft* v. *Staffordshire County Council* it fell to the Court of Appeal to decide whether partnership rearrangements gave rise to a sale, lease, transfer by inheritance, or transaction of comparable legal effect.[104] On 20 March 1989 Mr Victor Holdcroft, who was then farming in partnership with his wife and son, gave notice to terminate the tenancy which he held from Staffordshire County Council. The property was a dairy farm with quota. On 27 February 1990 his son and daughter-in-law purchased a new farm, without quota. Shortly afterwards, on 1 March 1990, a partnership reorganization took place, Mr and Mrs Victor Holdcroft retiring and their son entering into partnership with his wife (this partnership having a different name but the same milk quota registration number). The herd was moved to the recently purchased farm and milked by the son and his wife, while Mr Victor Holdcroft traded for a while on his own account on the land held from Staffordshire County Council prior to giving up possession.

The Court of Appeal, *inter alia*, felt that there was a change of occupation on the reorganization of the partnership, and that the change of occupation had comparable legal effect to a sale, lease, or transfer by inheritance. In this connection the Court was of the view that 'when the regulations talk about "transfer" they are talking in terms of occupation and not of legal title'.[105] However, the judgment did not address directly the effect of the United Kingdom provisions which barred the transfer of quota in the event of certain forms of transaction, including the grant of licences. As has been seen, these provisions may have particular relevance in the context of partnership reorganizations.[106] For example, since the coming into force of

[101] The Court criticized the fact that the SLOM had been made to the two partners rather than to the partnership.

[102] Case 86/90, [1992] ECR I–6251, [1993] 1 CMLR 489.

[103] See also Case 98/91, *Herbrink* v. *Minister van Landbouw, Natuurbeheer en Visserij*, [1994] ECR I–223, [1994] 3 CMLR 645.

[104] [1994] 28 EG 131.

[105] Ibid., at 134. Reference was made to the definitions of 'transferee' and 'transferor' in the Dairy Produce Quotas Regs. 1989, S.I.1989 No.380, reg.2(1) (retained in the Dairy Produce Quotas Regs. 1994, S.I.1994 No.672). See also *R.* v. *Ministry of Agriculture, Fisheries and Food*, ex p. *Cox*, [1993] 22 EG 111, [1993] 2 CMLR 917.

[106] See definition of 'producer' in Ch.1, 5.

the Dairy Produce Quotas (Amendment) Regulations 1988,[107] where a
landowning partner sharing occupation with another for the purposes of the
partnership business terminates the partnership and revokes that licence,
the transaction would not seem to satisfy the necessary criteria for the
transfer of the milk quota. Likewise, no transfer would appear to be triggered where the landowner enters into a further partnership with a new
partner and for that purpose grants a licence to share occupation. Similar
considerations apply upon the commencement and termination of share-
farming agreements, equally understood to be carried into effect through
the medium of licences. Indeed, it is generally a key purpose of such
transactions that the non-landowning partner or the share-farmer secures
no interest in the landowner's quota. None the less, notwithstanding the
United Kingdom provisions, there may be benefit in giving due weight to
the approach of the European Court as exemplified in such cases as *Von
Deetzen II*. Most notably, there is a willingness to look beyond the precise
nature of the transaction to the commercial reality.

4.4. Contractual provisions

Farmers and their advisers have sought to mitigate the dangers inherent in
quota transfers by means of contractual provisions — but the efficacy of
certain of these provisions may be questioned. Thus, in *Walker* v. *Titterton*
counsel conceded that the legal effect of the transfer rules could not be
overridden by a declaration in the contract, a concession which the judge
found helpful.[108] There would seem to be less objection to indemnities
against, for example, a claim on the quota made by the transferor's successor in title. It could be argued that such clauses seek to regulate the consequences of non-compliance with the transfer rules rather than to oust them
altogether — although it could be countered that where the clause is invoked the net effect would be similar.[109] In any event they undoubtedly
suffer from the usual weakness of indemnities, not least that the transferor
may be bankrupt or may have failed in turn to extract an indemnity from his
successor.

4.5. Re-identification of the producer's holding

Finally, it may be noted that there existed for some time a facility to
'manage' wholesale quota through re-identification of the producer's holding.[110] The provisions in question permitted such re-identification by agree-

---

[107] S.I.1988 No.534.   [108] (unreported, Stoke-on-Trent County Court, 23 Dec. 1993).

[109] Much will depend on broad policy issues regarding the extent to which contracting out is
permissible. See, in the context of the AHA 1986, e.g., *Johnson* v. *Moreton*, [1980] AC 37,
[1978] 3 All ER 37, [1978] 3 WLR 538.

[110] Council Reg. (EEC) 857/84, [1984] OJ L90/13, Art.8, as implemented in the UK by the
Dairy Produce Quotas Regs. 1986, S.I.1986 No.470, reg.11(1) (subsequently amended by the
Dairy Produce Quotas (Amendment) Regs. 1988, S.I.1988 No.534, reg.6).

ment between the purchaser and producer where the producer moved from one holding to another within the same region. A Form T3 was employed and the consent of all interested parties was required. However, in *Walker* v. *Titterton* it was held that the use of the re-identification procedure could not of itself effect a transfer of quota, which requires rather the transfer of a holding or part of a holding, accompanied by a change of occupation.[111] The same approach was also adopted in *W. E. & R. A. Holdcroft* v. *Staffordshire County Council*, a Form T3 being felt inappropriate to transfer the quota to the recently purchased farm.[112] As indicated, the re-identification procedure is no longer available. It was not re-enacted in the Dairy Produce Quotas Regulations 1989[113] and has not re-appeared in any subsequent legislation.

5. EXCEPTIONS TO THE GENERAL PRINCIPLE

While it has remained the general principle throughout that milk quota is linked to the holding, this general principle has been qualified by several important exceptions. The link, therefore, is not indissoluble; and perhaps it is this characteristic as much as any other which renders the nature of milk quota so elusive. The exceptions are, however, carefully circumscribed; and, even when upholding their efficacy, the European Court has also taken the opportunity to reassert the primacy of the general principle. Moreover, each exception may be viewed as directed to achieving specific objectives, most notably to mitigate the full effect of the general principle where it would work harshly upon certain categories of producers, or to promote flexibility where beneficial to the operation of the milk quota system as a whole.

Six exceptions may be considered.[114] First, as already seen, Council Regulation (EEC) 857/84 itself provided that, where there was the sale, lease, or transfer by inheritance of an undertaking, Member States could at their discretion add a part of the reference quantity to the national reserve;[115] and, under the current legislation, Article 7(1) of Council Regulation

---

[111] (unreported, Stoke-on-Trent County Court, 23 Dec. 1993).   [112] [1994] 28 EG 131.
[113] S.I.1989 No.380.
[114] The treatment of the various exceptions is not exhaustive. Rather, particular weight is attached to those of relevance to the milk quota system as operated in the UK. For consideration of exceptions, generally, see e.g., Laffoy, M., 'How are Milk Quotas to be characterised in Irish Land Law? Milk Quotas as Security for Loans', in *Milk Quotas: Law and Practice: Papers from the ICEL Conference — June 1989*, at 27–33. For judicial consideration, see, e.g., *Re the Küchenhof Farm*, [1990] 2 CMLR 289 (a decision of the German Administrative Court); *Faulks* v. *Faulks*, [1992] 15 EG 82; *Cottle* v. *Coldicott*, [1995] SpC 40; and *Harries* v. *Barclays Bank plc* (not yet reported, High Court, 20 Dec. 1995).
[115] [1984] OJ L90/13, Art.7(3). Clarification was supplied that wholesale quota would pass to the wholesale quota national reserve and direct sales quota to the direct sales quota national reserve: Council Reg. (EEC) 590/85, [1985] OJ L68/1. Subsequently Member States received a discretion to vary the amount added to the national reserve, in accordance with criteria

(EEC) 3950/92 states that: 'Any part of the reference quantity which is not transferred with the holding shall be added to the national reserve.'[116] It has also been seen that no 'siphon' to fuel the national reserve has been implemented in the United Kingdom, despite calls for such a measure as a means of making available quota for new entrants.[117]

It may be argued that this exception was complemented or extended by the provisions in the initial legislation permitting Member States, on the transfer of part of a holding, to disapply apportionment of quota where the area of land concerned fell below a minimum figure.[118] As indicated, this discretion was for some time exercised in the United Kingdom, there being no apportionment on a 'minor change of occupation'.[119] Later amendment permitted Member States the option to add the reference quantity attaching to the minimum area to the national reserve.[120] However, again no such provisions are currently implemented in the United Kingdom.[121]

Secondly, as also seen, quota could be removed from the holding to the national reserve under the various outgoers' schemes, in return for compensation. The provisions contained in Council Regulation (EEC) 857/84[122] have now been replaced by similar provisions in Council Regulation (EEC) 3950/92;[123] but, there is currently no outgoers' scheme operative in the United Kingdom, a significant factor no doubt being the high prices available on the open market for farmers ceasing milk production.

A third, and very important, exception was introduced explicitly to alleviate one of the situations where the general principle resulted in 'difficult situations at an economic and social level'.[124] It sought to reduce the hardship suffered by lessees who, on termination of their leases, sought to continue milk production elsewhere. Under the general principle the milk quota would revert to the landowner. To counter this, as early as 1985 a further exception was introduced by the insertion of a new Article 7(4) into Council Regulation 857/84.[125] Member States were granted the option to place all or part of the reference quantity at the disposal of the departing

relating to the size of the holdings concerned: Comm. Reg. (EEC) 1211/87, [1987] OJ L115/30; and Comm. Reg. (EEC) 1681/87, [1987] OJ L157/11.

[116] [1992] OJ L405/1.   [117] See Ch.4, 3.

[118] Comm. Reg. (EEC) 1371/84, [1984] OJ L132/11, Art.5(2).

[119] For first implementation in the UK, see the Dairy Produce Quotas Regs. 1984, S.I.1984 No.1047, reg.2(1) and Sch.1, Part III, and Sch.2, Part III; and Ch.4, 3.

[120] Comm. Reg. (EEC) 1546/88, [1988] OJ L139/12, Art.7(2).

[121] The discretion ceased to be exercised as from the Dairy Produce Quotas Regs. 1989, S.I.1989 No.380.

[122] [1984] OJ L90/13, Art.4(1)(a) and (2).   [123] [1992] OJ L405/1, Art.8.

[124] Council Reg. (EEC) 590/85, [1985] OJ 68/1, Preamble.

[125] [1984] OJ L90/13, the new Art.7(4) being inserted by Council Reg. (EEC) 590/85, [1985] OJ L68/1. For the detailed rules, see Comm. Reg. (EEC) 1043/85, [1985] OJ L112/18. See also Comm. Reg. (EEC) 1546/88, [1988] OJ L139/12, Art.7(4).

lessee, provided that he intended to continue milk production. Subsequently, the interaction of the general principle and this exception was considered in detail by the European Court in *Wachauf* v. *Bundesamt für Ernährung und Forstwirtschaft*;[126] and, whilst reaffirming that, as a rule, the reference quantity would return to the lessor who retook posession of the holding, the Court did emphasize that each Member State enjoyed a 'power' to allocate all or part of the reference quantity to the departing lessee. Indeed, the existence of this power contributed to the overall validity of the Community provisions governing the termination of leases: when it was taken together with the ability to compensate tenants wishing to discontinue milk production definitively,[127] Member States enjoyed a sufficiently wide margin of appreciation to enable them to apply those provisions consistently with the protection of the lessee's fundamental rights.[128] The United Kingdom has not exercised this discretion, preferring instead a detailed compensation scheme as provided under the Agriculture Act 1986.[129]

The effect of Article 7(4) of Council Regulation 857/84 has also been considered with regard to SLOM quota in the case of *Herbrink* v. *Minister van Landbouw, Natuurbeheer en Visserij*.[130] The facts were not simple. The applicant, a Dutch farmer, had as tenant entered into a non-marketing undertaking which expired after the introduction of milk quotas. He had resumed milk production in 1986, continuing that production until the expiration of his lease in 1987. The following year, 1988, he had again resumed milk production — but on another holding leased by an association which he had formed with, among others, his son-in-law. The European Court commenced by reaffirming that SLOM quota was equally subject to the general principle that every reference quantity is to remain attached to the land in respect of which it is allocated. None the less, where a Member State had opted to provide for allocations to departing lessees, the Court held that SLOM producers enjoyed the same rights as other producers, since they could expect to be able to carry on their activities under conditions which involved no discrimination between themselves and those other producers. The Netherlands having opted to exercise Article 7(4), the initial question before the Court was answered in the applicant's favour. A further question related to the effect of a change in business structure in these circumstances and, in particular, whether the fact that the applicant was no

---

[126] Case 5/88, [1989] ECR 2609, [1991] 1 CMLR 328.
[127] Under Council Reg. (EEC) 857/84, [1984] OJ L90/13, Art.4(1)(a).
[128] There is an argument that the lessee's fundamental rights were prejudiced, the operation of both Arts.4(1)(a) and 7(4) of Council Reg. (EEC) 857/84 being dependent on the exercise of the Member State's discretion. Cf., Case 120/86, *Mulder I*, [1988] ECR 2321, [1989] 2 CMLR 1.
[129] See Ch.5, 4.2.   [130] Case 98/91, [1994] ECR I–223, [1994] 3 CMLR 645.

longer *sole* lessee would prejudice his position. The Court held that an allocation should not be made if the association had been formed only for the purpose for realizing the marketable value of the SLOM quota for the benefit of the applicant. Otherwise, there was nothing in the regulations which precluded entitlement, the association as 'producer' being the correct recipient.[131]

Although Article 7(4) has been repealed, the potential hardship of the general rule on departing lessees is still specifically addressed, albeit in somewhat modified form. Article 7(2) of Council Regulation (EEC) 3950/92 reconfirms the primacy of the general rule stating that, in the absence of agreement between the parties, where rural leases are due to expire without the possibility of renewal on similar terms (or in situations involving comparable legal effects), 'the reference quantities available on the holdings in question shall be transferred in whole or in part to the producers taking them over, in accordance with provisions adopted or to be adopted by Member States'. However, in these circumstances account must be taken 'of the legitimate interests of the parties'.[132] Accordingly, the current legislation would still seem to confer upon individual Member States a degree of flexibility. In the case of tenants qualifying under the Agriculture Act 1986 the United Kingdom would seem to have already taken proper account of their interests. Although the reference quantities may be transferred to the incoming producer in accordance with the general rule, they benefit from an extensive statutory scheme of compensation. More recently, the ATA 1995 has introduced a further statutory scheme to cover farm business tenants.

Departing lessees were not the only category of producers who derived benefit from the 1985 amendments to Article 7 of Council Regulation 857/84. In addition, an exception was created in favour of producers whose land was transferred to the public authorities and/or for public use.[133] In these circumstances Member States enjoyed the option to provide that all or part of the reference quantity corresponding to the holding, or to the part of the

---

[131] This judgment is consistent with the Court's general approach on business reorganizations, as considered *supra*. Not least, it follows closely the judgment in Case 44/89, *Von Deetzen II*, [1991] ECR II–5119, [1994] 2 CMLR 487.

[132] [1992] OJ L405/1.

[133] Council Reg. (EEC) 857/84, [1984] OJ L90/13, Art.7(1), as inserted by Council Reg. (EEC) 590/85, [1985] OJ L68/1. See also Art.7(3), as inserted by Council Reg. (EEC) 590/85, which provided that in the case of the transfer of land to the public authorities and/or for public use, Member States were permitted to require that the corresponding reference quantity be assigned in full to the wholesale quota or direct sales quota national reserve. This discretion (which had the effect of preventing reference quantities being 'lost' or, at least, unutilized on a permanent basis) was not exercised in the UK. It too may be regarded as a break in the link between quota and the holding. For the detailed rules, see Comm. Reg. (EEC) 1043/85, [1985] OJ L112/18.

holding transferred, be put at the disposal of the departing producer, conditional upon the producer intending to continue milk production.[134] As with the provisions in favour of departing lessees, the requirement that the producer intend to continue production once more highlighted the determination of the legislature that milk quotas be utilized for the purposes of an active dairy business. This exception has also survived to the present day, again in somewhat modified form — Article 7(1) of Council Regulation (EEC) 3950/92 stipulates that 'where land is transferred to public authorities and/or for use in the public interest, or where the transfer is carried out for non-agricultural purposes, Member States shall provide that the measures necessary to protect the legitimate interests of the parties are implemented, and in particular that the departing producer is in a position to continue milk production if such is his intention'. The United Kingdom statutory instruments passed subsequent to Council Regulation (EEC) 3950/92 remain silent in this respect.

A fifth exception is the temporary transfer of milk quota during the currency of a milk year, commonly referred to as 'quota leasing'.[135] In the light of the sheer number and value of transactions, this is ensured a significant place in the operation of the milk quota system. Further, following early hostility from the Community institutions, its position has now become entrenched as an acceptable means of quota transfer, not least in that it provides a high degree of flexibility for producers.[136]

Such transfers, free from the requirement of a land transaction, were first expressly authorized by Council Regulation (EEC) 2998/87.[137] This regulation, which took effect as from the 1986/7 milk year, permitted the temporary transfer of that part of a producer's reference quantity which he did not intend to use. Further, it removed the prevailing uncertainty surrounding the form of transaction which had been conducted in the United Kingdom as a means of 'quota management'. Under Regulation 11 of the Dairy Produce Quotas Regulations 1986 producers had been permitted to trans-

---

[134] This option was not exercised in the UK.

[135] Both the Community regulations and the UK statutory instruments refer to 'temporary transfers' as opposed to 'leasing'. See, e.g., Council Reg. (EEC) 3950/92, [1992] OJ L405/1, Art.6; and the Dairy Produce Quotas Regs. 1994, S.I.1994 No.672, reg.15, as amended by the Dairy Produce Quotas (Amendment) (No.2) Regs. 1994, S.I.1994 No.2919. On temporary transfers, generally, see, e.g., Court of Auditors, *Special Report No.4/93 on the Implementation of the Quota System intended to control Milk Production*, [1994] OJ C12/1, at 4.48–50 (including a useful comparative discussion); Gregory, M. and Sydenham. A., *Essential Law for Landowners and Farmers*, at 120–1; Gehrke, H., *The Implementation of the EC Milk Quota Regulations in British, French and German Law*, at 71; Lennon, A. A. and Mackay, R. E. O. (edd.), *Agricultural Law, Tax and Finance*, at F2.4.

[136] For hostility to temporary transfers, see, e.g., *Court of Auditors' Special Report No.2/87 on the Quota/Additional Levy System in the Milk Sector*, [1987] OJ C266/1, at 4.20.

[137] [1987] OJ L285/1.

fer wholesale quota on a temporary basis.[138] Article 8 of Council Regulation
(EEC) 857/84 was regarded as authority for such transactions, in that it
permitted Member States operating Formula B to take the necessary steps
making it possible for purchasers 'to manage' their reference quantities.[139]

Notwithstanding early hostility, temporary transfers survived the reforms. Indeed, they came to be perceived as 'an improvement' to the
scheme and were correspondingly extended.[140] The rationale behind this
change in approach may be found in *Special Report No.4/93* of the Court of
Auditors, where temporary transfers were advocated in preference to *ex
post facto* offsetting. In particular, the former make producers more responsible for their production, in that, first, they reduce the amount of unutilized
reference quantities available for offsetting; and, secondly, the profitability
of additional production is decreased by the considerable payments
necessary to secure such temporary transfers.[141]

Under the current Community legislation, Article 6 of Council Regulation (EEC) 3950/92 provides that by 31 December, at the latest, Member
States are to authorize for that milk year temporary transfers of reference
quantities which producers do not intend to use. However, Member States
enjoy discretion as to the manner of implementation and, in any event, may
decide to refrain entirely on the basis of structural development and
adjustment and/or overriding administrative need. In the United Kingdom
implementing regulations the scope for temporary transfers has been concomitantly extended. Under Regulation 14 of the Dairy Produce Quotas
Regulations 1993 producers were permitted to transfer temporarily all,
rather than part, of their wholesale reference quantity to a producer or
producers supplying the same purchaser;[142] and, by further amendment
under Regulation 15 of the Dairy Produce Quotas Regulations 1994 the
derogation was applied as a general rule to any agreement between producers for the temporary transfer of reference quantities.[143] Accordingly, the
restriction to wholesale quota was removed, as also the requirement that
both the 'lessor' and 'lessee' supply the same purchaser. With the abolition
of the milk marketing boards and resulting increase in the number of
purchasers, this latter amendment should ensure that producers are not
prejudiced by the fact that their particular purchaser suffers from an absence of reference quantities available for temporary transfer.

[138] S.I.1986 No.470.
[139] [1984] OJ L90/13. On this aspect, see Wood, D. *et al.*, *Milk Quotas: Law and Practice*, at
15–16; and Milk Marketing Board, *Five Years of Milk Quotas: a Progress Report*, at 8.
[140] Council Reg. (EEC) 3950/92, [1992] OJ L405/1, Preamble.
[141] *Special Report No.4/93 on the Implementation of the Quota System intended to control
Milk Production*, [1994] OJ C12/1, at 4.50.
[142] S.I.1993 No.923. For the position prior to the reforms, see the Dairy Produce Quotas
Regs. 1991, S.I.1991 No.2232, reg.17(1).
[143] S.I.1994 No.672. See also *MAFF News Release* 44/94.

That having been said, there do remain certain limitations. Thus, temporary transfers are not permitted where they would increase or decrease the total direct sales or total wholesale quota for use by dairy enterprises within a Scottish Islands area.[144] Further, where a producer converts his reference quantity from wholesale quota to direct sales quota, or vice versa, he is not permitted to transfer quota in that milk year, whether on a temporary or permanent basis.[145]

A sixth exception was introduced by Council Regulation (EEC) 3950/92, which, in addition to extending temporary transfers of quota, provided the opportunity for permanent transfers of quota without land or of land without quota. While potentially critical in importance, the full legal ramifications of these provisions remain to be explored — and their practical effect as yet would appear limited. Thus, it has been provisionally calculated that in the 1994/5 milk year only 11 million litres had been so transferred throughout the whole United Kingdom.[146]

The Community regulation states that: 'With a view to completing restructuring of milk production at national, regional or collection area level, or to environmental improvement', Member States enjoy a discretion, *inter alia*, to authorize 'the transfer of reference quantities without transfer of the corresponding land, or vice versa, with the aim of improving the structure of milk production at the level of the holding or to allow for extensification of production'.[147] Moreover, Member States can provide, in the case of land transferred with a view to improving the environment, that the reference quantity available on the holding be allocated to the departing producer if he intends to continue milk production. In the United Kingdom, following consultation, Regulation 13 of the Dairy Produce Quotas Regulations 1994 authorized the transfer of quota without transfer of the corresponding land, it being very much in line with the Government's objective of 'decoupling' quota transfers from land transactions.[148] However, the terms upon which such transfers can occur are circumscribed by the Community legislation and, in particular, the need to demonstrate that genuine structural or environmental improvement will ensue. Furthermore, the United Kingdom

[144] The Dairy Produce Quotas Regs. 1994, S.I.1994 No.672, reg.15(4).

[145] The Dairy Produce Quotas Regs. 1994, S.I.1994 No.672, reg.18(4). Temporary (and permanent) transfers of quota are also restricted where a producer has been party to a 'transfer of quota without transfer of land' under the Dairy Produce Quotas Regs. 1994, S.I.1994 No.672, reg.13. Moreover SLOM 3 quota may not be the subject of a temporary transfer until 31 Dec. 1997: Council Reg. (EEC) 2055/93, [1993] OJ L187/8, Art.4.

[146] *Third Report from the Agriculture Committee: Trading of Milk Quota* (Session 1994–5, H.C.512), at 47 (figures supplied by MAFF).

[147] [1992] OJ L405/1, Art.8.

[148] S.I.1994 No.672. See also the Consultation Paper issued by MAFF, *Milk Quotas: Proposals to amend the Dairy Produce Quotas Regulations 1993*, at paras.12–17; and *MAFF News Releases* 328/93 and 44/94. For amendment of reg.13, see the Dairy Produce Quotas (Amendment) (No.2) Regs. 1994, S.I.1994 No.2919.

Government shared the Community institutions' determination to deter speculation. Accordingly, in the light of the numerous and detailed restrictions imposed by the Dairy Produce Quotas Regulations 1994, it is perhaps no surprise that permanent transfers of quota without land remain as yet a very small proportion of the total trade in quota.

The various constraints largely find expression within the application form itself (it being the obligation of the transferee to submit this form to the Intervention Board for approval at least ten working days prior to the intended transfer). These constraints include, first, a statement signed by both the transferor and transferee that they have agreed to the transfer, together with an explanation as to the reason why it is necessary to improve the structure of the business of the transferor or transferee; secondly, an undertaking by the transferor that he has not transferred reference quantities onto the holding under Regulation 13 in either the current or preceding milk year and that he will not transfer quota onto his holding (whether under Regulation 13 or otherwise) between the submission of the application and the end of the following milk year.[149] It is of note that the transferor must also undertake that he will not seek to circumvent these restrictions through connection with or involvement in another business. Likewise, the transferee must undertake that he will not transfer quota from his holding between the submission of the application and the end of the following milk year;[150] and that he will not seek to circumvent the restriction.[151]

However, in the case of the transferee's undertaking not to transfer quota, he may be released when, in the view of the Intervention Board, such release is justified by exceptional circumstances resulting in a significant fall in milk production which could not have been avoided or foreseen at the time of the submission of the application. The exceptional circumstances, listed in Regulation 13(8), are: the inability of the transferee to conduct his business over a prolonged period as a result of the onset of ill health, injury, or disability; a natural disaster seriously affecting the holding; the accidental destruction of buildings used for milk production; an outbreak of illness or disease seriously affecting the dairy herd; the serving of a notice or the making of a declaration under an order pursuant to section 17 of the Animal Health Act 1981[152] or the adoption of an emergency order under section 1 of the Food and Environment Protection Act 1985; the loss of a significant proportion of the forage area through compulsory purchase of all or part of the holding; and, where the transferee is a tenant, the serving of an incontestable notice to quit under section 26 of and Schedule 3 to the AHA 1986 (dealing with the serving of incontestable notices to quit where

---

[149] For these latter purposes a temporary transfer would amount to a breach of the undertaking, but not a transfer by inheritance: S.I.1994 No.672, reg.13(3).

[150] Again a temporary transfer would amount to a breach, but not a transfer by inheritance: ibid.

[151] S.I.1994 No.672, reg.13(2).   [152] Special provisions apply for Northern Ireland.

the tenant is not at fault). Again special rules apply to dairy enterprises in a Scottish Islands area, with no such transfer to be approved where it would result in an increase or reduction of total direct sales or wholesale quota available for use in that area.[153] However, while Regulation 13 does permit transfers of quota without land, the United Kingdom has not taken advantage of the opportunity afforded by the Community legislation to authorize transfers of land without quota.

Finally, it may be reiterated that the principle under which quota is linked to the holding receives modification in the case of SLOM quota, the purpose of the restriction being to prevent producers disposing of their allocations for a windfall profit.[154] As already indicated, in the case of SLOM 1 quota the sale or lease of the holding before 31 March 1992 led to forfeiture of the allocation to the Community reserve;[155] and in the case of SLOM 2 quota forfeiture to the national reserve was incurred on the sale or lease of the holding prior to 1 July 1994.[156] Since those dates are now past, the holdings in question may be sold or leased with impunity; and, moreover, temporary transfers of SLOM 2 quota have been permitted since 31 March 1995.[157] That having been said, sales or leases of the holding before 1 October 1996 will still result in the forfeiture of SLOM 3 quota to the national reserve; and, where only part of the holding is sold or leased, the amount forfeited is calculated on the basis of the area comprised in the transaction. Further, the temporary transfer of SLOM 3 quota between producers is ineffective until 31 December 1997.[158]

The detailed United Kingdom provisions are contained in Regulation 8 of the Dairy Produce Quotas Regulations 1994.[159] These provide, *inter alia*, that the proportion of SLOM 3 quota to be returned to the national reserve on the sale or lease of part of the holding shall be the same proportion which the agricultural area of the holding sold or leased bears to the total agricultural area farmed by the producer. In accordance with the general rules governing transfer, forfeiture will not occur on the grant of a licence or, in England and Wales, the grant of a tenancy for a period of less than ten months. Again in line with the Community provisions, there is also no forfeiture where the transfer of the holding occurs on inheritance, by way of gift for no consideration, or, broadly, where a tenancy passes by succession under the AHA 1986.[160] However, in all these cases the person to whom the

---

[153] S.I.1994 No.672, reg.13(9).
[154] See, in particular, Council Reg. (EEC) 764/89, [1989] OJ L84/2, Preamble; and Case 44/89, *Von Deetzen II*, [1991] ECR I–5119, [1994] 2 CMLR 487.
[155] Council Reg. (EEC) 764/89, [1989] OJ L84/2. By later amendment it was provided that forfeiture would be to the national reserve.
[156] Council Reg. (EEC) 1639/91, [1991] OJ L150/35.
[157] Council Reg. (EEC) 3950/92, [1992] OJ L405/1, Art.6(1).
[158] Council Reg. (EEC) 2055/93, [1993] OJ L187/8, Art.4.   [159] S.I.1994 No.672.
[160] In this connection the detailed provisions preclude forfeiture: (1) on the grant of a tenancy following a direction under s.39 or s.53 of the AHA 1986 to a successor on the death or retirement of the previous tenant; (2) on the grant of a new tenancy by agreement to a

holding or part of it is transferred must undertake to comply with requirements in relation to SLOM quota undertaken by his predecessor.

There are, accordingly, many exceptions to the general principle. None the less, as perhaps most clearly stated in *Herbrink* v. *Minister van Landbouw, Natuurbeheer en Visserij*, the general principle remains in place, enshrined in both Community and national legislation;[161] and the exceptions may be perceived rather as specific measures to address specific objectives which cannot be achieved by application of the general principle. For that reason, it may difficult to establish a consistent thread running through the various exceptions. For example, temporary transfers are directed to ensuring efficient and flexible management of reference quantities. In particular, as highlighted by the Court of Auditors, it has the effect of limiting the adverse consequences of offsetting. By contrast, the provision under which Member States could permit certain departing tenants to retain their reference quantities may be seen as a response to an identified 'difficult situation'. With such differing objectives, it may be seen as inevitable that the nature of quota and, more specifically, the nature of the link with the holding should prove so elusive.

6. APPORTIONMENT AND AREAS USED FOR MILK PRODUCTION

On the sale, lease, or transfer by inheritance of part of a holding, as already mentioned, it becomes necessary to apportion the milk quota between the transferor and transferee.[162] The initial leglislation provided that apportionment was to be made between 'the producers operating the holding in proportion to the areas used for milk production or according to other objective criteria laid down by Member States'.[163] In implementing this provision, the United Kingdom soon showed a willingness to accept apportionment 'according to areas used for milk production' rather than laying down other objective criteria.[164] Such a basis survived largely unchanged

person entitled following a direction under s.39 of the AHA 1986 (within s.45(6) of the AHA 1986); (3) on the grant of a tenancy by agreement to a close relative under s.39(1)(b) or (2) of the AHA 1986; or (4) on any other grant of a tenancy by a landlord to a successor of a tenant who has died or retired.

[161] Case 98/91, [1994] ECR I–223, [1994] 3 CMLR 645.

[162] On these aspects, generally, see, e.g., Wood, D. *et al.*, *Milk Quotas: Law and Practice*, at 18–22; Rodgers, C. P., *Agricultural Law*, at 319–20; Gregory, M. and Sydenham, A., *Essential Law for Landowners and Farmers*, at 117–9; Sir Crispin Agnew of Lochnaw Bt, 'Apportionment of Milk Quota', [1992] Journal of the Law Society of Scotland 29–32; and Gehrke, H., *The Implementation of the EC Milk Quota Regulations in British, French and German Law*, at 72–3.

[163] Comm. Reg. (EEC) 1371/84, [1984] OJ L132/11, Art.5(2).

[164] The Dairy Produce Quotas Regs. 1984, S.I.1984 No.1047, Sch.1, para.19 (in respect of direct sales quota); and Sch.2, para.19 (in respect of wholesale quota); the Dairy Produce

until the reforms, at which point greater flexibility was introduced. Article 7 of Council Regulation (EEC) 3950/92 provided that reference quantities available on a holding are to be transferred with the holding on sale, lease, or transfer by inheritance 'to the producers taking it over in accordance with detailed rules to be determined by the Member States taking account of the areas used for dairy production or other objective criteria and, where applicable, of any agreement between the parties'. Certain differences in the wording may be highlighted. First, rather than apportionment 'in proportion to' areas used for milk production or 'according to' any other objective criteria, it is required only that these be taken into account. Secondly, the Community legislation introduces specific reference to 'any agreement between the parties' (where applicable).[165] However, it would not seem that such agreement will be conclusive. Again it appears no more than a factor to be taken into account; and the wording would not suggest that it precludes the need also to consider the areas used for milk production or any other objective criteria.

For England and Wales these provisions may now be found implemented by Regulations 7–12 of and Schedule 2 to the Dairy Produce Quotas Regulations 1994.[166] Where there is agreement and this is set out in the prescribed notice, the agreement will form the basis of the apportionment. That having been said, both the transferor and the transferee must confirm that it takes account of the areas used for milk production. Further, it is open to the Intervention Board to require arbitration to determine the correct apportionment where it has reasonable grounds for believing that the areas used for milk production are not as specified in the notice; or that the areas used for milk production were not as agreed between the parties at the time of apportionment (notwithstanding that no such notice was submitted).[167] In particular, the milk marketing boards, and now the Intervention Board, have sought to prevent the transfer of an unjustifiable amount of milk quota per hectare or acre (commonly known as 'quota loading'). For these purposes the apportionment of over 20,000 litres per hectare or 8,094 litres per acre has given rise to closer investigation.[168]

A point of great importance is that, should the apportionment be determined by arbitration (either because no agreement has been reached or

---

Quotas (Amendment) Regs. 1985, S.I.1985 No.509; and the Dairy Produce Quotas Regs. 1986, S.I.1986 No.470, Sch.4.

[165] It had earlier been established that leaving apportionment to agreement between the parties would not qualify as laying down objective criteria: Case 121/90, *Posthumus* v. *Oosterwoud*, [1991] ECR I–5833, [1992] 2 CMLR 336.

[166] S.I.1994 No.672. With regard to Scotland and Northern Ireland, see respectively Sch.3 and Sch.4.

[167] Ibid., reg.12.

[168] This limit has been stipulated in guidance notes; and see too *Hansard* (HL) Vol.458, Col.777.

following such referral by the Intervention Board), the arbitrator is to base his award 'on findings made by him as to areas used for milk production in the last five-year period during which production took place before the change of occupation'.[169] This wording was first enacted in the Dairy Produce Quotas Regulations 1993[170] which contained a significant change from the Dairy Produce Quotas Regulations 1991.[171] Under the 1991 Regulations the arbitrator was directed to base his award on findings made by him as to areas used for milk production in the five years preceding the change of occupation rather than, as now, the last five-year period during which production took place. This takes into account producers who have temporarily transferred their quota, or who have suffered quota confiscation.[172]

Accordingly, even following the increase in flexibility conferred by the reforms, the 'areas used for milk production' remain central to the question of apportionment.[173] Neither the Community nor United Kingdom regulations contain a statutory definition; but the courts have adopted a broad interpretation. Early authority was provided by the United Kingdom High Court decision of *Puncknowle Farms Ltd.* v. *Kane* decided under the Dairy Produce Quotas Regulations 1984.[174] Restriction to land used for the current production of milk was rejected. Rather 'areas used for milk produc-

---

[169] The Dairy Produce Quotas Regs. 1994, S.I.1994 No.672, Sch.2, para.3(1) (in the case of England and Wales).

[170] S.I.1993 No.923.   [171] S.I.1991 No.2232.

[172] Such confiscation may also be the occasion of an apportionment: the Dairy Produce Quotas Regs. 1994, S.I.1994 No.672, reg.33(5)(b), as amended by the Dairy Produce Quotas (Amendment) Regs. 1994, S.I.1994 No.2448, and (in the case of England and Wales) Sch.2, paras.1(5) and 3(2)).

[173] It may be noted that the current Community provisions refer to 'dairy' rather than 'milk' production: Council Reg. (EEC) 3950/92, [1992] OJ L405/1, Art.7. Further, 'areas used for milk production' is but one of many similar expressions found in the milk quota legislation. It may be contrasted, for example, with land used 'for the feeding of dairy cows kept on the land' (Agriculture Act 1986, Sch.1, para.6(1)(a)); and land used 'for the feeding, accommodation or milking of dairy cows kept on the land' (Agriculture Act 1986, Sch.1, para.7(1)(b)). Both of these arise in the context of a tenant's claim for compensation under the Agriculture Act 1986, considered in Ch.5, 4.2. The SLOM regulations has thrown up further alternatives. It has been seen that, where part of a holding was sold during the course of a non-marketing or conversion undertaking, any allocation of SLOM quota fell to be apportioned: Case 81/91, *Twijnstra* v. *Minister van Landbouw, Natuurbeheer en Visserij*, [1993] ECR I–2455; and in these circumstances the apportionment must be made in proportion to the 'areas under forage': Council Reg. (EEC) 2055/93, [1993] OJ L187/8, Art.1(2). Other examples are found in the provisions which govern forfeiture of SLOM quota. Thus, SLOM 1 quota was to be forfeited, on the sale or lease of part of the holding before 31 Mar. 1992, 'on the basis of the feed-crop area sold or leased': Council Reg. (EEC) 764/89, [1989] OJ L84/2; SLOM 2 quota was to be forfeited, on the sale or lease of part of the holding before 1 July 1994, 'on the basis of the fodder area sold or leased': Council Reg. (EEC) 1639/91, [1991] OJ L150/35; and, as seen, under the United Kingdom legislation SLOM 3 quota is to be forfeited, on the sale or lease of a holding before 1 Oct. 1996, in 'the same proportion which the agricultural area of the holding sold or leased bears to the total agricultural area farmed by the producer': the Dairy Produce Quotas Regs. 1994, S.I.1994 No.672, reg.8(3).

[174] [1985] 3 All ER 790, (1985) 275 EG 1283, [1986] 1 CMLR 27.

tion' included also 'areas which are used to support the dairy herd by the maintenance of animals between one lactation and another, and to support the animals which are destined for inclusion in the dairy herd on any holding'.[175] Further, land used for milk production may be used concomitantly for other purposes, such as the shooting of game. The judge also specifically accepted that the buildings and yards of a dairy unit would be included (if of sufficient size to be material), as would land used for dairy or dual-purpose bulls (if bred to enter the production herd and not for sale).

However, an aspect which remained in doubt was whether the milk quota should be apportioned evenly across all areas used for milk production, on a mathematical basis; or whether, adopting an 'artistic approach', a greater amount should be apportioned to those parts of the holding apprehended to make a greater contribution to milk production. On the latter basis, for example, rich grazing close to the dairy itself would receive a greater amount of quota than marginal land serving as rough grazing for young stock. In *Puncknowle Farms Ltd.* v. *Kane* this issue was not the subject of detailed consideration, it being conceded that the stocking rate was not a relevant factor. That having been said, while the Irish High Court in *Lawlor* v. *Minister for Agriculture* again accepted a broad interpretation of 'areas used for milk production' (including expressly land occupied by replacement calves), there was distinct preference for the artistic approach.[176] In particular, when addressing the transfer or subdivision of a holding between the national reference year and the introduction of the milk quota system, it was the view of the Court that the persons acquiring such holdings should acquire also reference quantities appropriate to the amount and the *nature* of the lands.

The question came before the European Court in the case of *Posthumus* v. *Oosterwoud*, where the United Kingdom, intervening, argued that the mathematical approach should not necessarily be applied.[177] The Court laid emphasis on the fact that Member States enjoyed an option to lay down objective criteria; and these could, and indeed should, relate to the characteristics of the holding or of the farming activities. Thus, a Member State had not laid down objective criteria where the national legislation provided for apportionment to be carried out primarily in accordance with the terms agreed between the parties. Further, in the absence of objective criteria, apportionment was to be strictly in accordance with the areas used for milk production. As a result in this event no account could be taken of the extent to which the different areas contributed to total milk production, so ensuring both legal certainty and the effectiveness of the scheme.[178]

---

[175] Ibid., at 794 (All ER), 1284 (EG), and 33 (CMLR).

[176] [1990] 1 IR 356, [1988] IRLM 400, [1988] 3 CMLR 22.

[177] Case 121/90, [1991] ECR I–5833, [1992] 2 CMLR 336.

[178] Those parts of the decision which deal with buildings on the holding are not easy to interpret — as later observed by the Advocate-General in Case 79/91, *Knüfer* v. *Buchmann*, [1992] ECR I–6895, [1993] 1 CMLR 692. That having been said, it would seem to have been the

This position was confirmed in *Knüfer* v. *Buchmann*, where the European Court upheld both the broad interpretation of areas used for milk production and the mathematical approach to apportionment (in default of Member States enacting objective criteria to displace such a basis).[179] Most notably, it was held that for the purposes of apportionment 'all the areas of the holding which contribute directly or indirectly to milk production must be taken into account, including the farmyard, the buildings and areas of roads serving the holding, provided that they contribute directly or indirectly to the milk production of the holding'.[180]

The United Kingdom having enacted no objective criteria to the contrary, it would seem that milk quota is to be apportioned on a mathematical basis, with no account taken of the relative contribution of different areas to milk production. While, as highlighted by the European Court in *Posthumus* v. *Oosterwoud*, this does have the advantage of legal certainty, it is arguable that inconsistencies arise. For example, if the transferor's holding comprises, on the one hand, 50 acres of rich grassland upon which the milking herd is grazed, together with the milking facilities, and, on the other hand, 450 acres of marginal outlying land used to rear dairy replacements, it might seem surprising that 90 per cent of the milk quota attached to the latter.

In this context, the parties may have recourse to prospective apportionment, a facility first introduced under the Dairy Produce Quota Regulations 1986.[181] Under the current legislation, the Dairy Produce Quotas Regulations 1994, the occupier of the holding concerned may apply to the Intervention Board on the prescribed form, requesting either that a prospective apportionment be made taking account of areas used for milk production as set out in the application, or that a prospective apportionment be ascertained by arbitration. Again the arbitrator is directed to base his award on findings made by him as to areas used for milk production. With regard to prospective apportionments, the relevant period for him to conduct these findings is the last five-year period during which production took place before his appointment.[182]

The advantage of a prospective apportionment, whether by agreement or arbitration, is that it provides a degree of certainty for proposed land transactions. On a change of occupation of part of a holding during the

view of the Court that the land on which the buildings stand is to be taken into account, but is to receive the same weighting as ordinary grazing land (unless Member States enact objective criteria otherwise).

[179] [1992] ECR I–6895, [1993] 1 CMLR 692.
[180] Ibid., at I–6916 (ECR) and 705 (CMLR).
[181] S.I.1986 No.470, reg.8(4) and Sch.4, Part II.
[182] S.I.1994 No.672, reg.11 and (in the case of England and Wales) Sch.2. The prescribed form is MQ/8; and the request may be revoked by notice in writing to the Intervention Board, signed by the occupier of the holding to which the prospective apportionment relates.

ensuing period prescribed by the regulations, it will govern the apportionment of quota — and this is so even if the prospective apportionment is not
yet completed. In the case of prospective apportionments by agreement, the
request must have been made within the six months preceding the change
of occupation; and the occupier must have submitted a notice indicating
that the apportionment has been agreed. In the case of prospective
apportionments by arbitration, the change of occupation must occur within
the six months following the arbitration. As with apportionments in general, the Intervention Board requires arbitration if it has reasonable
grounds for believing that the areas used for milk production are not as
specified or as were agreed.[183]

In England and Wales any arbitration, whether relating to apportionment on transfer or to a prospective apportionment, is conducted for the
most part according to the procedure under the AHA 1986.[184] Thus, for
example, the arbitrator must in general make and sign his award within
fifty-six days of his appointment. However, there are certain differences.
Most notably, the AHA 1986 procedure governs disputes between landlords and tenants, whereas any person with an interest in the holding may
make representations to the arbitrator under the Dairy Produce Quotas
Regulations.[185] Accordingly, not just landlords and tenants may be involved, but also, for example, mortgagees.[186] Moreover, the dispute may not
concern a landlord or tenant at all, arising rather between two freehold
owners.

Two consequences of these apportionment provisions may be highlighted. First, they affect the way in which milk quota 'attaches' to the
holding. As indicated, the areas used for milk production remain of central
importance, not least since they form the basis of any arbitration award.
Accordingly, it may be argued that milk quota attaches only to those parts
of the holding which are used for milk production; and that it is capable of
moving from one part of the holding to another over a period of time
depending upon land use. However, in practice this may only become
critical during those five years of milk production immediately preceding

---

[183] Ibid., reg.12.

[184] For the detailed rules in the case of England and Wales, see ibid., Sch.2. For those under
the AHA 1986, see the AHA 1986, Sch.11; and also, for general discussion, e.g., Muir Watt, J.,
Agricultural Holdings, at 273–317; Densham, H. A. C., *Scammell and Densham's Law of
Agricultural Holdings*, at 363–86; and Rodgers, C. P., *Agricultural Law*, at 359–75. For the
procedure in Scotland, see the Dairy Produce Quotas Regs. 1994, Sch.3; and for the procedure
in Northern Ireland, see ibid., Sch.4. It may be noted that in England and Wales the original
legislation provided for arbitration under the Arbitration Act 1950. The introduction of a
procedure similar to that under the AHA 1986 was effected by the Dairy Produce Quotas
Regs. 1989, S.I.1989 No.380. However, the Arbitration Acts 1950–1979 have become very
relevant in the agricultural context, since (subject to exceptions) they govern disputes under
the 1995 Act.

[185] The Dairy Produce Quotas Regs. 1994, S.I.1994 No.672, Sch.2, para.16.

[186] For 'parties having an interest in the holding', see Ch.4, 7.1.1.

any change of occupation of a part of the holding. Until 'crystallized' by such an event, the need to identify the areas concerned is less pressing.[187] That having been said, a producer may be able to exploit these provisions at any time for purposes of estate management. In particular, if his holding comprises both freehold and tenanted land, (provided the tenancy agreement so permits) there should be the opportunity to concentrate milk production on the freehold land. Then, to the extent that the quota attaches to the freehold land, he should be able to realize its full value (as opposed, in the case of tenanted land, to sharing in that value through, for example, the receipt of compensation).[188]

Secondly, for all these reasons it is of paramount importance that at any time producers should have available records sufficient to establish the areas used for milk production. While in many cases the collation of this evidence should not give rise to difficulties (for example, the location of the milking parlour itself), there are circumstances in which ascertaining the relevant details may prove more onerous. For example, it is necessary to distinguish between, on the one hand, that part of the holding used for growing corn to be fed to the dairy cows, heifers, and other replacements, and, on the other hand, that part of the holding used for growing corn to be sold off the farm.

7. FURTHER ASPECTS OF QUOTA TRANSFERS

7.1. Permanent transfers

In addition to the general rules underlying permanent quota transfers, certain more practical aspects may be addressed. In this context five matters may be highlighted: the protection accorded to interested parties other than the transferor and transferee, for example landlords and mortgagees; the effect of the transfer on the butterfat base of the transferor and transferee; the effect of quota use during the year of transfer; the obligations imposed on purchasers; and the importance of complying with deadlines.

7.1.1. Third party interests

In the United Kingdom protection for third parties has long been afforded by the requirement that those third parties give their consent on the prescribed transfer form.[189] Statutory authority for this requirement was

---

[187] This aspect is highlighted by the Special Commissioners in *Cottle* v. *Coldicott*, [1995] SpC 40.

[188] For more detailed consideration of landlord and tenant issues, see Ch.5.

[189] In the case of permanent transfers with land the prescribed form is MQ/1. In the case of permanent transfers without land the prescribed form is MQ/2.

provided on first implementation of the milk quota system;[190] and the current provisions are contained in Regulations 7 and 13 of the Dairy Produce Quotas Regulations 1994, the former applying in the case of permanent transfers with land, the latter in the case of permanent transfers without land.[191] In either case the transferor may be in a position to provide a notice that he enjoys a 'sole interest' in the entirety of the holding; but, if this is not so, he should obtain the requisite consents of other interested parties.[192] Like provisions apply where a prospective apportionment is requested.[193] The more detailed contents of this 'consent or sole interest notice' may be found in the definition set out in regulation 2(1): the appropriate person must certify '(a) either that he is the occupier of that holding or part of a holding and that no other person has an interest in that holding or part of the holding, or (b) that all persons having an interest in the holding or part of the holding the value of which interest might be reduced by the apportionment or prospective apportionment to which the notice relates agree to that apportionment or proposed prospective apportionment'.

The terms of this definition must be broad enough to embrace landlords, since, as seen, in the United Kingdom the quota reverts to the landlord on the termination of a tenancy, subject to the right of certain tenants to claim compensation.[194] Moreover, if the transferor's holding comprises both freehold and tenanted land, the consent of the landlord should be obtained notwithstanding that the parties to the transfer may apprehend that the quota transferred is attached to the freehold element. That having been said, this requirement may perhaps be disputed in an extreme case. If all milk production has been conducted on the freehold land for the last five years, it is at least arguable that the tenanted land will not constitute a part of the 'holding' — and that, accordingly, landlord's consent need not be obtained.[195]

[190] The Dairy Produce Quotas Regs. 1984, S.I.1984 No.1047. See, in particular, the definition of 'consent or sole interest notice' in reg.2(1) and (in respect of direct sales quota) Sch.1, Part III, and (in respect of wholesale quota) Sch.2, Part III.

[191] S.I.1994 No.672.

[192] Together with the consent or sole interest notice there should be the statement, signed by the transferor and transferee, that they have agreed that the quota shall be apportioned taking account of the areas used for milk production as specified in the statement.

[193] S.I.1994 No.672, reg.11(3).

[194] However, for a somewhat surprising view taken by the German Administrative Court, see *Re the Küchenhof Farm*, [1990] 2 CMLR 289.

[195] See the definition of 'holding' considered in Ch.1, 5; and also the views of Professor Usher as followed by the County Court in *Carson* v. *Cornwall County Council*, [1993] 03 EG 119. The safer course must be to obtain consent — with recourse to arbitration as a sanction if it is not forthcoming. It may also be mentioned that, where there is more than one landlord (e.g., the producer effecting the transfer has taken several short-term tenancies), consent should be obtained from each landlord, notwithstanding that the change of occupation which triggers the transfer of quota affects only a part of the Euro-holding.

In the case of mortgagees or chargees there are again vexed issues.[196] The Intervention Board form for permanent transfers of quota clearly envisages that mortgagees have an interest in the holding; and there can be no doubt that any such interest would be materially devalued if the quota were permanently transferred to the holding of another producer. However, preliminary questions arise as to the extent that milk quota may be the subject of a mortgage or charge at all. In this regard it may be argued that quota is caught by a mortgage under the Law of Property Act 1925, on the basis that it constitutes an interest in land; or, alternatively, it may be seen as 'farming stock' or 'other agricultural assets', so falling within the scope of the Agricultural Credits Act 1928.

Turning first to mortgages under the Law of Property Act 1925, it is certain is that those drafting the 1925 legislation could never have been expected to address such matters; and farmers and their advisers are again confronted by considerable problems in fitting the novel 'asset' constituted by milk quota into pre-existing property legislation. However, it may be emphasized that one of the rights enjoyed by a legal mortgagee is the right to possession; and, while subject to restrictions, a legal mortgagee exercising that right may physically move onto the land.[197] That this action triggers a change of occupation for the purposes of the milk quota legislation has now been confirmed in *Harries* v. *Barclays Bank plc*.[198] By virtue of its possession the Bank was held entitled to become the registered producer. For the same reason the Bank was also held entitled to all the proceeds of sale of the farm in question, when subsequently sold with the benefit of quota. Indeed, entitlement extended to the proceeds of temporary transfers made by the Bank pending sale, insofar as those proceeds were apportionable to the farm. Accordingly, possession proved the critical factor, the Court considering it unnecessary to determine whether quota is an asset in its own right. However, there was some doubt whether the quota fell subject

---

[196] For general discussion of milk quota and charges, see, e.g., Laffoy, M., 'How are Milk Quotas to be characterised in Irish Land Law? Milk Quotas as Security for Loans', in *Milk Quotas: Law and Practice: Papers from the ICEL Conference — June 1989*, at 27–33; and Anderson, H, *Agricultural Charges and Receivership* (Chancery Law Publishing, London, 1992), at 21–2.

[197] The Law of Property Act 1925, ss.85(1), 86(1), and 87(1). Under these provisions a legal mortgage must be effected by lease or, as is more usual, by charge; and in the latter case the mortgagee enjoys the same protection, powers, and remedies as if he had taken a lease. As a result, strictly speaking, the mortgagee 'may go into possession before the ink is dry on the mortgage unless there is something in the contract, express or by implication, whereby he has contracted himself out of that right': *Four-Maids Ltd.* v. *Dudley Marshall (Properties) Ltd.*, [1957] Ch.317, per Harman J, at 320. However, in practice the right is now heavily circumscribed: see, e.g., Burn, E. H., *Cheshire and Burn's Modern Law of Real Property* (15th edn., Butterworths, London, 1994), at 691–3.

[198] (not yet reported, High Court, 20 Dec. 1995). There must be considerable uncertainty as to whether the mere right to possession triggers a change of occupation: see, e.g., *R.* v. *Ministry of Agriculture, Fisheries and Food*, ex p. *Cox*, [1993] 22 EG 111, [1993] 2 CMLR 917.

to the legal charge. At the same time a purposive approach dispensed with the argument that, since it was neither selling nor supplying milk or milk products, the Bank could not quality as a producer so as to effect temporary transfers.

Had the decision, been that the quota was a separate asset of the mortgagor, the extent of mortgagees' security would have been diminished;[199] and even mortagors would still experience difficulties. For example, if the mortgagee went into possession (and, more specifically, took physical occupation), the mortgagor would seem to be confronted by severe obstacles when seeking to dispose of the quota to a third party.[200]

A case where the bank apparently had no charge over the quota was *Huish* v. *Ellis*. However, the mortgagee took the practical course of selling the quota with the land and the Court held that such action would not give rise to liability, provided that the mortgagee acted in good faith.[201] In particular, there was no breach of the duty to secure the best price reasonably obtainable — which, according to the mortgagor, would have resulted from separate sales of the land and the quota.

In any event, the mortgagee may be exposed to significant dangers in protecting his security during the currency of the mortgage. For example, many deeds will not contain express provisions which address the preservation of quota; and prior to 1 September 1995 the AHA 1986 provided that the mortgagee could not as a general rule exclude the right of the mortgagor to create leases over the mortgaged agricultural land.[202] Accordingly, mortgagors could enter into the tenancies of ten months or more necessary to effect permanent transfers of quota and the tenancies would be binding

---

[199] Mortgagees who had lent money on the security of land prior to the introduction of milk quotas could have found themselves prejudiced by a switch in value from the land to the quota, notwithstanding that they had neither knowledge nor expectation of such an eventuality at the time of the mortgage.

[200] By way of illustration, under the general rule he would be required to enter into a land transaction in respect of the land occupied by the mortgagee. If the mortgagee became the registered producer by virtue of possession (notwithstanding that the quota were treated as the separate asset of the mortgagor), presumably the mortgagee would hold the quota on trust for the mortgagor.

[201] [1995] NPC 3.

[202] Under s.99 of the Law of Property Act 1925 a mortgagor can grant certain forms of lease over the mortgaged land so as to bind the mortgagee. However, by virtue of s.99(13) this power could be excluded by contrary intention expressed in the mortgage deed or otherwise in writing; and in practice such a clause has been frequently inserted in the mortgage deed. Subsequently, the AHA 1986 provided that the mortagor's ability to lease could not be excluded in relation to mortgages of agricultural land granted after 1 Mar. 1948 (agricultural land being defined for these purposes as in the Agriculture Act 1947, s.109): s.100 and Sch.14, para.12. For a commentary, see Densham. H. A. C., *Scammell and Densham's Law of Agricultural Holdings, 1993 Supplement*, at A22–4. For limitations on such leases of agricultural land, see *Agricultural Mortgage Corporation plc* v. *Woodward*, [1995] 04 EG 143 (transaction set aside under Insolvency Act 1986, s.423). For express cousideration of a mortgagor's ability to dispose of quota by lease, see *Harries* v. *Barclays Bank plc* (not yet reported, High Court, 20 Dec. 1995).

on the mortgagee. However, the 1995 Act now ensures that, subject to certain exceptions, any mortgage made on or after 1 September 1995 can include terms to prevent the creation of leases over the mortgaged agricultural land.[203] This should prove a potent addition to the arsenal of mortgagees.

With regard to charges under the Agricultural Credits Act 1928, it would seem that milk quotas may only be captured by such charges if they constitute 'farming stock' or 'other agricultural assets'.[204] Besides, agricultural charges may only be given by farmers in favour of banks. Considering first the definition of 'farming stock', this would appear exhaustive; and none of the matters listed match quotas of any form.[205] As for the definition of 'other agricultural assets', prior to the 1995 Act these were narrowly defined as 'a tenant's right to compensation under the Agricultural Holdings Act 1986, except under section 60(2)(b) or 62, for improvements, damage by game, disturbance or otherwise, and any other tenant right'. For this reason 'agricultural assets' may only be charged by a tenant farmer. That having been said, there is an argument that compensation payable to departing tenants in respect of milk quota under the Agricul-ture Act 1986 would qualify as 'any other tenant right'; and the definition has been expanded by the 1995 Act in a limited, yet significant, manner so as to include a tenant's right to compensation under section 16 of that Act (which is taken to cover compensation in respect of milk quota).[206]

Accordingly, there remains considerable conceptual uncertainty as to the interest of mortgagees in milk quota; and these conceptual uncertainties are exacerbated by the practical difficulties in preserving and realizing any security which may properly be taken. As mentioned, this state of affairs may be seen as an almost inevitable consequence of any attempt to reconcile milk quota with a property law developed with no such 'asset' in mind. None the less, the questions which arise are of critical importance, going to the very foundations of lending to the dairy sector.

---

[203] S.31. The exceptions are set out in the Agricultural Tenancies Act 1995, s.4; and include successions under the AHA 1986 and variations of AHA 1986 Act tenancies.

[204] S.5(7).

[205] 'Farming stock' is defined as 'crops or horticultural produce, whether growing or severed from the land, and after severance whether subjected to any treatment or process of manufacture or not; live stock, including poultry and bees, and the produce and progeny thereof; any other agricultural or horticultural produce whether subjected to any treatment or process of manufacture or not; seeds and manures; agricultural vehicles, machinery, and other plant; agricultural tenant's fixtures and other agricultural fixtures which a tenant is by law authorised to remove'. It may be noted that in *Harries* v. *Barclays Bank plc* the parties agreed that nothing turned on the terms of the agricultural charges: (not yet reported, High Court, 20 Dec. 1995).

[206] See the 1995 Act, Sch., para.7; and e.g., Anderson, H., *Agricultural Charges and Receivership*, at 21–2. On compensation payable to departing tenants in respect of milk quota, generally, see Ch.5, 4.

Finally, as indicated, interested parties are entitled to make representations at any arbitration.[207] Since it may be difficult to ascertain which parties do have an interest, there would seem merit in such matters being dealt with as a preliminary issue. Support for the arbitrator having authority to do so may be gleaned from the approach of the Court of Appeal in *W. E. & R.A. Holdcroft* v. *Staffordshire County Council*, where there was a willingness to accord him broad jurisdiction.[208]

7.1.2. *The effect on the butterfat base*

The effect of permanent transfers on the butterfat base need only be considered in the case of wholesale quota, the butterfat rules being inapplicable to direct sales quota. However, as indicated, a small percentage rise or fall in the butterfat base may have severe consequences and, indeed, may prove the decisive factor when determining whether a producer has exceeded his individual reference quantity. In the light of such consequences, the Community legislation lays down detailed provisions which address the treatment of the butterfat base in the context of permanent transfers.[209] Under Commission Regulation (EEC) 536/93 it is now provided that in the case of such transfers the butterfat base is transferred with the reference quantity with which it is associated. Where the transferee has no existing quota, his butterfat base is that of the quota transferred; but if he already has quota registered in his name, then the butterfat base of that existing quota must be adjusted. The calculation is carried out by ascertaining the average butterfat base of the transferred and the existing quota, as weighted by the respective amounts of the transferred and the existing quota. If the transferor retains any quota, its butterfat base is not affected.[210]

7.1.3. *Quota use*

As the system is currently operated in the United Kingdom, the parties are to agree the amounts of the transferred quota which are respectively used and unused during that milk year. Used quota then remains registered with

[207] The Dairy Produce Quotas Regs. 1994, S.I.1994, No.672, (in the case of England and Wales) Sch.2, para.16.

[208] [1994] 28 EG 131.

[209] For these purposes permanent transfers include transfers which may be authorized by Member States under Council Reg. (EEC) 3950/92, [1992] OJ L405/1, Art.8 — i.e., in the UK transfers of quota without land as authorized by the Dairy Produce Quotas Regs. 1994, S.I.1994 No.672, as amended by the Dairy Produce Quotas (Amendment) (No.2) Regs. 1994, S.I.1994 No.2919.

[210] [1993] OJ L57/12, Art.2(1)(c). For a useful worked example, see the Explanatory Notes accompanying forms MQ/1 and MQ/2.

the transferor's purchaser until the commencement of the next year, at which point it becomes available for the transferee. For these purposes proper account must be taken of any wholesale deliveries and/or direct sales already made, with any appropriate butterfat adjustment in the case of wholesale quota. Failure to reach agreement results in the respective amounts of used and unused quota being determined mandatorily on the basis of wholesale deliveries and/or direct sales up to the date of the transfer.[211]

7.1.4. Purchasers

Until recently the effect of transfers of wholesale quota had little effect on the reference quantities of the purchasers to whom the milk was delivered. First, there was a high chance that both the transferor and transferee supplied the same purchaser in view of the dominant role of the five milk marketing boards; and, secondly, there was no scope for transfers between producers in different regions. Circumstances have now materially changed, in that the milk marketing schemes have been abolished, throwing open the arena to a wider variety of purchasers;[212] and, since the enactment of the Dairy Produce Quotas Regulations 1993, there has been free mobility of quota throughout the United Kingdom, with the exception of dairy enterprises located within a Scottish Islands area.[213] Accordingly, although there were provisions governing such adjustments prior to the changes, the legislation in this regard has now assumed far greater practical importance.[214]

The current Community rules address the adjustment of purchasers' reference quantities where a purchaser in whole or part replaces one or more purchasers; and where a producer transfers from one purchaser to another. In both of these circumstances the reference quantities available to producers shall be taken into account for the remainder of the milk year, less any deliveries already made. Account is also taken of the requisite butterfat content.[215] These rules are now implemented in Regulation 6 of

---

[211] See the Explanatory Notes accompanying forms MQ/1 and MQ/2.

[212] Under the Agriculture Act 1993 and, in the case of Northern Ireland, the Agriculture (Northern Ireland) Order 1993, S.I.1993 No.2665 (N.I.10).

[213] S.I.1993, No.923, reg.7(7)(f). See now the Dairy Produce Quotas Regs. 1994, S.I.1994 No.672, reg.7(7). For the rules governing regions immediately prior to the Dairy Produce Quotas Regs. 1993, see the Dairy Produce Quotas Regs. 1991, S.I.1991 No.2232, reg.6.

[214] Indeed, as from the inception of the milk quota system legislation addressed replacement of purchasers and the effect of transfers between producers on purchasers: Council Reg. (EEC) 857/84, [1984] OJ L90/13, Art.7(2); Commission Reg. (EEC) 1371/84, [1984] OJ L132/11, Art.6(1) (c) and (d); and the Dairy Produce Quotas Regs. 1984, S.I.1984 No.1047, Sch.3.

[215] Council Reg. (EEC) 3950/92, [1992] OJ L405/1, Art.2(2).

the Dairy Produce Quotas Regulations 1994, laying down a detailed procedure and demanding deadlines.[216]

7.1.5. Deadlines

In order to facilitate the smooth operation of the milk quota system, there
are a number of critical deadlines to be met by producers effecting permanent transfers of milk quota. As already indicated, the calculation of any
superlevy payable cannot in any event be completed until some time after
the end of the milk year in question; and without deadlines the period of
uncertainty would be yet further extended. In this context certain provisions may be highlighted.[217] First, any producer permanently transferring
wholesale quota (whether with or without land) must notify the purchaser
with whom his quota is registered within seven working days of the transfer.[218] Secondly, in the case of a permanent transfer with land the transferee
must submit the appropriate documentation to the Intervention Board
within twenty-eight days of the change of occupation. Moreover, the deadline is tightened where the change of occupation takes place towards the
close of the milk year. In any event the documentation must be submitted
no later than seven working days after the end of the milk year; and failure
to meet this requirement has severe consequences. For the purpose of the
superlevy calculation the unused quota transferred does not count as part of
the transferee's entitlement for the milk year in which the transfer took
place, instead becoming available for reallocation; and the transferee cannot demand that by reason of the transfer an amendment be made to the
the amount of quota, if any, reallocated to him for the milk year in which
the transfer took place.[219] Thirdly, in the case of a permanent transfer
without land, the application form must be submitted to the Intervention
Board by the transferee no later than ten working days before the intended
date of such transfer;[220] and, where the Intervention Board approves the
transfer, the transferee must within twenty-eight days from the transfer, and
in any event no later than seven working days from the end of the milk year
in question, submit a statement of the amounts of used and unused quota

---

[216] S.I.1994 No.672, reg.6, as amended by the Dairy Produce Quotas (Amendment) Regs.
1994, S.I.1994 No.2448 (in consequence of the reorganization of milk marketing in Great
Britain), and the Dairy Produce Quotas (Amendment) Regs. 1995, S.I.1995 No.254 (in consequence of the reorganization of milk marketing in Northern Ireland). For treatment of deliveries already made during a milk year in which a producer changes from one purchaser to
another, see ibid., reg.6(4).
[217] For the tightening of deadlines, generally, see Commission Reg. (EEC) 536/93, [1993] OJ
L57/12; and Ch.3, 2.2.4.
[218] The Dairy Produce Quotas Regs. 1994, S.I.1994 No.672, reg.6(5).
[219] Ibid., reg.7(1) and (3).   [220] Ibid., reg.13(1).

available to the transferor and transferee as at the date of the transfer. Failure to meet this last requirement results in revocation of Intervention Board approval.[221] Finally, in the case of a proposed transfer of all or part of a holding by a producer with SLOM quota still subject to restrictions, the transferor must submit the requisite notice of transfer to the Intervention Board at least twenty-eight days before the transfer, together with any other related evidence within such time as the Intervention Board might reasonably require.[222]

7.2. Temporary transfers

In the United Kingdom temporary transfers are also conducted through the Intervention Board by means of a prescribed form,[223] and a reasonable charge may be made by the Board.[224] The form does not refer to any requirement for landlord's or mortgagee's consent, nor is there any reference to consents in the Community or United Kingdom regulations. However, a landlord may wish to include a term to this effect in any tenancy agreement.[225] The effect on the butterfat base is the same as with permanent transfers, albeit for the limited period of a milk year.[226] With regard to quota use during the milk year of the temporary transfer, the Community regulations expressly provide that only quota which the producer does not intend to use may be so transferred; and the United Kingdom regulations, in slightly different terms, confine temporary transfers to 'unused quota'.[227] The requisite documentation must as a general rule be submitted no later than 15 December in the milk year of the transaction.[228] The deadline, imposed to prevent a surge of applications in the last quarter of the milk year, has historically been the source of difficulties; and once again difficulties were encountered in the 1994/5 milk year. So that producers could rearrange their affairs to take account of the revocation of the milk marketing schemes, it proved necessary to delay the last day of submission until 31 December (the last day permitted under the Community regulations).[229]

---

[221] Ibid., reg.13(5) and (6).   [222] Ibid., reg.9.

[223] The prescribed form is MQ/3 in the case of wholesale quota and MQ/4 in the case of direct sales quota.

[224] The charge is currently £17.63p.

[225] On this aspect, see further Ch.5, 2.

[226] Comm. Reg. (EEC) 536/93, [1993] OJ L57/12, Art.2(1)(c).

[227] Council Reg. (EEC) 3950/92, [1992] OJ L405/1, Art.6(1); and the Dairy Produce Quotas Regs. 1994, S.I.1994 No.672, reg.15(1).

[228] The Dairy Produce Quotas Regs. 1994, S.I.1994 No.672, reg.15(3).

[229] The Dairy Produce Quotas (Amendment) (No.2) Regs. 1994, S.I.1994 No.2919. See also *IB Press Notice* 15/94. For earlier difficulties, see *MAFF News Release* 438/93.

5

Landlord and Tenant Issues

1. GENERAL[1]

Since milk quotas, from their inception in 1984, have been critical to the profitable operation of dairy enterprises and, moreover, quickly acquired an economic value, there can be little surprise that they have thrown landlord and tenant issues into sharp relief.[2] Indeed, it may be seen as inevitable that tension should arise between the competing interests of the landlord, to whose land the milk quota is 'linked', and the tenant, who as a rule will be the occupier and registered producer. For example, a landlord will be eager to preserve that link. He will also share with the tenant concern as to the effect of quota on the rent — and this consideration is thrown into sharp focus where the tenant overtly turns the quota to account by means of temporary transfers. However, arguably the greatest source of conflict arises on the termination of the tenancy. In accordance with the general principle the quota in the United Kingdom reverts to the landlord or other incoming occupier and the tenant must look to compensation through statute or agreement for the protection of his interest.

Accordingly, in this context three aspects may be considered: first, the extent to which a landlord may entrench 'attachment' of the quota to

[1] It is throughout critical to distinguish between the Agricultural Holdings Act 1986 (which consolidated the law of agricultural holdings) and the Agriculture Act 1986 (which introduced a milk quota compensation scheme for tenants). The former will continue to be abbreviated to 'the AHA 1986', while the latter will be given its full title. The Agricultural Tenancies Act 1995 (which introduced the farm business tenancy régime) will continue to be abbreviated to 'the ATA 1995'.

[2] This area is very well served by existing literature. There is substantial coverage in leading treatments of the AHA 1986: see, e.g., Slatter, M. and Barr, W., *Farm Tenancies*, at 83–4 and 255–6; Muir Watt, J., *Agricultural Holdings*, at 56 and 225–42; Densham, H. A. C., *Scammell and Densham's Law of Agricultural Holdings*, at 97–9 and 318–35; Gregory, M. and Sydenham, A., *Essential Law for Landowners and Farmers*, at 121–6; Rodgers, C. P., *Agricultural Law*, at 65–6, 96–7, and 325–36; Gehrke, H., *The Implementation of the EC Milk Quota Regulations in British, French and German Law*, at 74–80; Hill and Redman's *Law of Landlord and Tenant*, at F224–40; and Woodfall's *Law of Landlord and Tenant* (looseleaf edn., Sweet and Maxwell, London), at 21.285–312. In addition, there are more specialist works written prior to the introduction of the farm business tenancy régime: see, e.g., Wood, D. *et al.*, *Milk Quotas: Law and Practice*, at 31–51; and Wood, D., Priday, C., Moss, J. R., and Carter, D., *The Handbook of Milk Quota Compensation* (Farmgate Communications, 1987), *passim*. For the position under the farm business tenancy régime, see, e.g., Evans, D., *The Agricultural Tenancies Act 1995*, at 8/15 and 16; and Sydenham, A. and Mainwaring, N., *Farm Business Tenancies: Agricultural Tenancies Act 1995*, at 40–1 and 70.

the land comprised in the tenancy by means of 'quota clauses'; secondly, the treatment of milk quota on rent review; and, thirdly, the treatment of milk quota on the termination of a tenancy.

A point of some importance is that landlord and tenant issues remain very much the province of the individual Member States, with the Community institutions showing a marked reluctance to intervene in this sphere. As the Advocate-General stated in *Wachauf* v. *Bundesamt für Ernährung und Forstwirtschaft*, such an approach is logical 'given the diversity of national legal systems and implementing legislation and the different circumstances of individual producers'.[3] Accordingly, as again stated by the Advocate-General in *Wachauf*, it is 'for the national court to determine in the concrete case whether and to what extent account should be taken of the tenant's interest in the quota'.[4] That having been said, the European Court may indicate in general terms the applicability of the general principles of Community law, the principles of non-discrimination and of respect for the right to property enjoying particular relevance. Confirmation of this stance may be found in the judgment of the European Court in *R.* v. *Ministry of Agriculture, Fisheries and Food*, ex parte *Bostock*, where it was affirmed that 'legal relations between lessees and lessors, in particular on the expiry of a lease, are, as Community law now stands, still governed by the law of the Member State in question'.[5] The very considerable discretion which the implementing legislation has accorded to Member States in this area may be regarded as consistent with such an approach.[6] For example, under Article 7(4) of Council Regulation (EEC) 857/84 Member States were free to choose whether or not the reference quantity should remain attached to the holding on the termination of a lease, or whether it should be made available to the departing lessee if he intended to continue milk production.[7] The relevant legislation now in force is Article 7(2) of Council Regulation (EEC) 3950/92, which stipulates that:

Where there is no agreement between the parties, in the case of rural leases due to expire without any possibility of renewal on similar terms, or in situations involving comparable legal effects, the reference quantities available on the holdings in question shall be transferred in whole or in part to the producers taking them over, in

---

[3] Case 5/88, [1989] ECR 2609, [1991] 1 CMLR 328, at 2628 (ECR) and 340 (CMLR).

[4] Ibid., at 2631 (ECR) and 344 (CMLR).

[5] Case 2/92, [1994] ECR I–955, [1994] 3 CMLR 547, at I–985 (ECR) and 571 (CMLR).

[6] For comparison of the tenant's status in the UK, France, and Germany, see Gehrke, H., *The Implementation of the EC Milk Quota Regulations in British, French and German Law*, at 145–8. For a decision of the German Administrative Court where the interest of the tenant was favoured as against that of the landlord, see *Re the Küchenhof Farm*, [1990] 2 CMLR 289. More specifically, the consent of the landlord was not required as a condition of the tenant's participation in a national outgoers' scheme.

[7] [1984] OJ L90/13, as inserted by Council Reg. (EEC) 590/85, [1985] OJ L68/1.

accordance with provisions adopted or to be adopted by the Member States, taking account of the legitimate interests of the parties.[8]

Accordingly, within the framework of the general principles of Community law there remains much scope to accommodate the disparate landlord and tenant legislation found throughout the Member States.

However, notwithstanding this latitude, the interaction of Community and national measures has on occasion given rise to difficulties. One clear example may be reiterated – the potential for confusion implicit in the expression 'holding'. As indicated, for the purposes of Community legislation this expression is currently defined as 'all the production units operated by the single producer and located within the geographical territory of a Member State'.[9] However, for the purposes of United Kingdom legislation a 'holding' has generally referred to the land comprised in a single agricultural tenancy agreement – which may be only one of several production units operated by a single producer.[10] It has also been seen that, for this reason, the term 'Euro-holding' is frequently employed to describe a holding under the milk quota legislation, in contradistinction to an agricultural holding within the AHA 1986, or the land comprised within a farm business tenancy under the ATA 1995.[11]

2. QUOTA CLAUSES

It has long been in the financial interests of a landlord to secure the attachment to his land of the greatest quantity of quota.[12] Indeed, since 25 September 1986 the Agriculture Act 1986 has given statutory recognition of such increase in value.[13] In pursuing this objective, a landlord will be confronted by several obstacles. Thus, the tenant may attempt the permanent disposal to another producer of the reference quantity attaching to the holding. That having been said, it is unlikely that landlords will experience this drastic step in practice. In the first place it has been seen that for the

---

[8] [1992] OJ L405/1. It may be noted that in these circumstances the general rule under which reference quantities are transferred with the holding would only seem to apply where there is no agreement between the parties.

[9] Council Reg. (EEC) 3950/92, [1992] OJ L405/1, Art.9(d), as amended by Council Reg. (EEC) 1560/93, [1993] OJ L154/30.

[10] See the AHA 1986, s.1; and the ATA 1995, s.38(1). See, generally, Ch.1, 5.

[11] For judicial concern lest the two be confused, see *Broadland Properties Estates Ltd.* v. *Mann*, [1994] SLT (Land Ct.) 7. A point of some importance is that for the purposes of the milk quota compensation provisions in the Agriculture Act 1986 recourse was had to the contemporary Community definition of 'holding' as opposed to the definition found in the agricultural holdings legislation.

[12] On quota clauses see, in particular, Wood, D. *et al., Milk Quotas: Law and Practice,* at 37–9; and Rodgers, C. P., *Agricultural Law,* at 65–6.

[13] The Agriculture Act 1986, Sch.1, para.9.

purposes of permanent transfers with or without land a landlord is an interested party whose consent must be obtained to any apportionment of quota. In default of such consent, the matter must be referred to arbitration.[14] Moreover, under the Community legislation it is stated that Member States implementing the option to permit quota transfers without the corresponding land should take into account 'the legitimate interests of the parties'.[15] Nevertheless, in the case of transfers with land perhaps the most potent weapon in the landlord's arsenal is the high probability that the tenant will be prohibited from entering into the appropriate form of land transaction. As indicated, in England and Wales there should be, at the least, the grant of a tenancy for a period of ten months; and in most tenancy agreements subletting is either prohibited absolutely or subject to landlord's consent.[16] Moreover, where subletting is subject to landlord's consent, there is no statutorily imposed requirement that such consent must not be unreasonably withheld — the operation of section 19 of the Landlord and Tenant Act 1927 being excluded in the case of agricultural holdings under the AHA 1986 and farm business tenancies under the ATA 1995.[17] Besides, if a tenancy agreement which does not contain a prohibition against subletting is governed by the AHA 1986, a landlord may at any time serve a request that one be incorporated.[18] Finally, tenancy agreements entered into after the introduction of milk quotas are in any event likely to contain an express provision which precludes the tenant from effecting permanent transfers of quota without the landlord's written consent.

Accordingly, a landlord enjoys considerable protection should his tenant seek to effect an outright permanent transfer to another producer of the reference quantity attaching to the tenanted land. That having been said, a landlord could easily be prejudiced by a less drastic but highly effective measure open to a tenant who farms a 'composite' holding — i.e., a

[14] Ch.4, 7.1.1. See, in particular, the Dairy Produce Quotas Regs. 1994, S.I.1994 No.672, regs.7(2)(b) and 13(2)(b). However, it is not clear from these provisions that the landlord enjoys any significant sanction where the apportionment is conducted correctly.

[15] Council Reg. (EEC) 3950/92, [1992] OJ L405/1, Art.8.

[16] Under the AHA 1986 breach of such a covenant would entitle the landlord to serve notice to quit, and possibly to rely on Case E in Sch.3 (on the ground that his interest had been materially prejudiced by an irremediable breach). See, e.g., *Snell* v. *Snell*, (1964) 191 EG 361; and, for a more benign view, *Pennell* v. *Payne*, [1995] QB 192, [1995] 2 All ER 592, [1995] 2 WLR 261, [1995] 06 EG 152. Difficult questions arise from the fact that, notwithstanding the breach of covenant, the subtenancy would remain valid until the head-tenancy was determined by the head-landlord (so bringing down the interest derived from it). Accordingly, on grant the subtenancy would seem capable of triggering a transfer of quota. However, if the subtenancy was brought down before 10 months had elapsed, it must be doubted whether the transfer would be fully effective; and the grantor could find himself open to litigation from both his landlord and his subtenant.

[17] See now the ATA 1995, Sch., para.6.

[18] The AHA 1986, s.6. For the immediate effectiveness of such a request, see s.6(5). Although this facility is not available under the farm business tenancy régime, most well drawn agreements would already contain the requisite express covenant barring subletting.

Euro-holding comprising both the land held under the tenancy and other land (for these purposes generally freehold land). The tenant with other freehold land could concentrate his dairy enterprise on that part of the Euro-holding rather than the land subject to the tenancy, with a view to securing that on any subsequent apportionment, taking account of the 'areas used for milk production', the majority (if not all) of the reference quantity is attached to the former. Indeed, if all dairy activities took place on the freehold land for a period of five years, then the tenant should be able to realize the full value of the milk quota;[19] and, besides, he should be able to dispose of it by the grant of a tenancy over the freehold land, unfettered by any covenant against subletting.

In seeking to prevent such stripping of quota from the tenanted land, a landlord may again have recourse to the terms of the tenancy agreement; and again agreements entered into after the inception of the milk quota system could be expected to contain clauses protecting his interest. However, such a 'quota clause' is not one that the landlord can request under section 6 of the AHA 1986;[20] nor does it receive any mention in the ATA 1995. Accordingly, its inclusion would be a matter for negotiation between the parties — and may need to be bought at the expense of concession on other items. At the same time, since such 'quota massage' would not in itself seem to give rise to a change of occupation and apportionment, a landlord could not take advantage of the requirement for his consent under the transfer regulations. Further, it is unlikely that the landlord will be able to seek a remedy based on breach of the rules of good husbandry found in section 11 of the Agriculture Act 1947. Although section 11 requires, *inter alia*, that the occupier maintain 'a reasonable standard of efficient production, as respects both the kind of produce and the quality and quantity thereof', there is no limitation to a particular form of production; and a shift from an efficient dairy enterprise to, for example, an efficient beef enterprise would not seem to constitute a breach.[21]

Thus the dangers faced by a landlord will, in all probability, be exacerbated where the tenancy was granted prior to 2 April 1984. Not only are such agreements most unlikely to contain covenants specifically directed to milk quotas; but the landlord's interests would not seem to be protected by

---

[19] The Dairy Produce Quotas Regs. 1994, S.I.1994 No.672, (in the case of England and Wales) Sch.2, para.3(1): an arbitrator is directed to base his award on 'findings made by him as to areas used for milk production in the last five-year period during which production took place before the change of occupation'.

[20] The matters for which provision is to be made in written tenancy agreements are set out in the AHA 1986, Sch.1.

[21] Wood, D. *et al.*, *Milk Quotas: Law and Practice*, at 38. However, for the view that the tenant was under a statutory duty to 'husband' and protect the milk quota so that it remained attached to the holding, see the unreported English County Court case of *Wilson* v. *Hereford and Worcester County Council*; and Densham, H. A. C., *Scammell and Densham's Law of Agricultural Holdings, 1993 Supplement*, at A12.

more general covenants relating to marketing schemes (such as the Potato
Marketing Scheme). In *Lee* v. *Heaton* the landlord enjoyed the benefit of a
covenant, *inter alia*, 'Not without the Landlord's consent in writing to
permit or suffer any basic quota under a marketing scheme to lapse or be
reduced through any failure to produce on the Holding a sufficient quantity
of any produce now or hereafter the subject of a marketing scheme which
is or has been produced on the Holding or which is normally grown in the
district'; and 'Not to dispose of the whole or any part of any basic quota
under a marketing scheme allotted to the Holding or to the Tenant in
respect of the Holding'.[22] The agreement being dated 24 May 1974, there
was no direct reference to milk quotas. The Dairy Produce Quota Tribunal
held that milk quota did not constitute basic quota under a marketing
scheme, the milk quota regulations imposing a levy on excess production
rather than establishing or seeking to establish a regulated market.

While quota clauses have demonstrated considerable variety in their
drafting and, in any event, are likely to be the subject of close negotiation
between the parties, there are certain matters which may generally be
expected. For example, where the farm is exclusively dairy by nature, the
tenant may be required to use it for that purpose alone. If, by contrast, the
farm is used for dairy purposes only in part, there may be an obligation not
to reduce the areas used for milk production. In any event a landlord will be
concerned to see the level of dairy production maintained.[23] Further, the
tenant is likely to be prohibited from disposing of or dealing in quota
apportioned, or which may be apportioned, to the tenanted land; and
another prevalent clause is that the tenant co-operate with the landlord on
the termination of the tenancy in transferring the milk quota to the landlord
or the landlord's nominee (the latter frequently being the incoming tenant).
At the same time there would be merit in identifying the areas used for milk
production and in imposing on the tenant an obligation to maintain records
of land use, for example details of the fields grazed by the dairy herd or
followers. Although this gives rise to a considerable administrative burden,
in the light of the value of milk quota such a provision may not be regarded
as unduly onerous.[24]

An express quota clause, once incorporated, should not only have the
advantage of addressing specific issues of concern to the landlord. It also
offers the opportunity for the landlord to seek an injunction where the

[22] (1987) 283 EG 1076.

[23] In an extreme case the absence of wholesale deliveries, direct sales, or temporary transfers would lead to confiscation of quota: Council Reg. (EEC) 3950/92, [1992] OJ L405/1, Art.5;
and the Dairy Produce Quotas Regs. 1994, S.I.1994 No.672, reg.33, as amended by the Dairy
Produce Quotas (Amendment) Regs. 1994, S.I.1994 No.2448. See Sydenham, A. and
Mainwaring, N., *Farm Business Tenancies: Agricultural Tenancies Act 1995*, at 41.

[24] See, e.g., Wood, D. *et al.*, *Milk Quotas: Law and Practice*, at 39; and Muir Watt, J.,
Agricultural Holdings, 1989 Supplement, at 11–12.

tenant is in breach or, more radically, to serve an effective notice to quit. In the latter case, where the tenancy is governed by the AHA 1986 it would be possible to argue that certain breaches at least are irremediable for the purposes of Case E in Schedule 3. If that ground is available, the tenant's only remedy is to contest the reasons stated in the notice to quit by demanding arbitration within one month of receipt of the notice. Alternatively, even if the breach is remediable, Case D in the same Schedule would seem to be available, with the result that, should the tenant fail to comply with the requisite notice requiring remedy of the breach within a reasonable time, again his only remedy will be to contest the reasons stated in the notice to quit.[25] Where the tenancy is governed by the ATA 1995, then the normal rules of forfeiture should apply.[26]

That having been said, the validity of express quota clauses may be open to challenge. Where the AHA 1986 applies, it could be alleged that many contravene section 15, which, subject to certain conditions, entitles the tenant without incurring any penalty, forfeiture, or liability, to dispose of the produce of the agricultural holding, other than manure produced on it; and also to practise any system of cropping of the arable land. While these provisions may not be excluded by the terms of the contract, the effectiveness of any challenge on this ground must be weakened by the fact that such freedom of cropping is confined to 'arable land'.[27] Moreover, the provision has not survived into the ATA 1995. Secondly, outgoers' schemes have raised particular difficulties.[28] For a tenant to take advantage of these schemes, the landlord's consent was required. The grounds for refusal were restricted but, if the application succeeded, the landlord was entitled to share in the cessation payments. The consent provisions were considered before the European Court in *Wachauf* v. *Bundesamt für Ernährung und Forstwirtschaft*.[29] In particular, the Advocate-General addressed the question whether the 1984 Outgoers' Scheme implemented in Germany gave rise to a breach of the general principles of Community law. On his analysis the need for consent was not, in principle, objectionable; but, if the landlord enjoyed an unqualified power of veto, this might constitute a breach of the

---

[25] For the operation of Cases D and E, see, e.g., Muir Watt, J., *Agricultural Holdings*, at 114–24; Densham, H. A. C., *Scammell and Densham's Law of Agricultural Holdings*, at 173–94; and Rodgers, C. P., *Agricultural Law*, at 130–40.

[26] The absence of the statutory procedure under Cases D and E may lead to significant delays.

[27] S.15(7) states that: '"arable land" does not include land in grass which, by the terms of a contract of tenancy, is to be retained in the same condition throughout the tenancy'.

[28] See, e.g., the Milk (Community Outgoers Scheme) (England and Wales) Regs. 1986, S.I.1986 No.1611, as amended by the Milk (Community Outgoers Scheme) (England and Wales) (Amendment) Regs. 1987, S.I.1987 No.410, and the Milk (Community Outgoers Scheme) (England and Wales) (Amendment) (No.2) Regs. 1987, S.I.1987 No.909; and the Milk (Cessation of Production) (England and Wales) Regs. 1987, S.I.1987 No.908.

[29] Case 5/88, [1989] ECR 2609, [1991] 1 CMLR 328.

principle of non-discrimination 'in that the same requirement would be applied to all tenant farmers irrespective of their individual situation and in particular of their contribution to the acquisition of the quota'.[30] However, following the reference to Luxembourg, the German Administrative Court held that the requirement for landlord's consent was void.[31] The ground for the decision was that it led to unequal treatment between owner-occupiers and lessees which was not objectively justifiable. Accordingly, it gave rise to a breach of the principle of equal treatment contained in Article 3(1) of the Constitution and of the prohibition against discrimination contained in Article 40(3) of the Treaty.[32] Thus, in at least one Member State the Court has shown an inclination to preserve a tenant's right to realize the value of milk quota.

Finally, in the case of temporary transfers by a tenant the Dairy Produce Quotas Regulations 1994 impose no requirement that a consent notice be submitted; and, moreover, such transfers can be effected without the need for any land transaction. For these reasons it would seem that a landlord would be well advised to insist upon an express provision either prohibiting the temporary transfer of quota absolutely or permitting it subject to his written consent.[33] There is Scottish authority to the effect that a tenant who enters into temporary transfers year after year may be in breach of the rules of good husbandry.[34] In particular, the Court emphasized that such a tenant may cease to be regarded as a producer.[35] That having been said, the presence of express provision would seem preferable to this more uncertain remedy.[36]

---

[30] Ibid., at 2629 (ECR) and 341 (CMLR).

[31] *Re the Küchenhof Farm*, [1990] 2 CMLR 289.

[32] Most notably, the Court was not persuaded that such consent was necessary to protect the landlord's 'expectation' of quota on the expiry of the lease; and, consequently, if the quota was released to the Government through the outgoers' scheme, the landlord's rights were not infringed. In support of this conclusion emphasis was laid on the connection between quota and production rather than ownership of the capital used to produce milk. The Court stated, *inter alia,* that 'there is no fundamental connection between the milk quota and ownership of land, but only between the quota and the milk output for the reference year': ibid., at 302.

[33] However, it must be recognized that such a provision could prove a significant factor on rent review.

[34] *Cambusmore Estate Trustees* v. *Little*, [1991] SLT (Land Ct.) 33 (considering the Agricultural Holdings (Scotland) Act 1948, Sch.6, rule 1). If the Agricultural Land Tribunal granted a certificate that the tenant was not complying with the rules of good husbandry, the landlord could serve notice to quit under (in the case of England and Wales) Case C of Sch.3 to the AHA 1986.

[35] This risk would now seem to be reduced, not least since temporary transfers of quota are sufficient to prevent confiscation under the Dairy Produce Quotas Regs. 1994, S.I.1994 No.672, reg.33.

[36] The legislation governing unfair contract terms may also be noted. The Unfair Contract Terms Act 1977 does not apply to contracts relating to the creation or transfer of interests in land, or the termination of such interests. Thus, the grant or termination of a tenancy is excluded. It is less clear whether such transactions are covered by Council Directive 93/13/EEC, [1993] OJ L95/29, as implemented in the UK by the Unfair Contract Terms in Consumer

3. RENT

In the case of tenancies protected by the AHA 1986, section 12 and Schedule 2 lay down a statutory procedure governing rent review.[37] As a rule such rents may be reviewed every three years and, where the parties cannot reach agreement, either party may demand arbitration. Should there be a demand for arbitration, the arbitrator is directed to assess 'the rent at which the holding might reasonably be expected to be let by a prudent and willing landlord to a prudent and willing tenant'. A point of some importance is that, if the agricultural holding subject to rent review is part of a larger Euro-holding under the Community legislation, it is first necessary to determine the amount of quota attaching to that agricultural holding.[38]

When assessing the rent the arbitrator is directed to take into account

Contracts Regs. 1994, S.I.1994 No.3159. Both the Community and national provisions only apply as against sellers of goods and suppliers of goods or services; and, while land contracts are not expressly excluded, very arguably their subject-matter is neither 'goods' nor 'services': see, e.g., Treitel, G. H., *The Law of Contract* (9th edn., Sweet and Maxwell, London, 1995), at 255; but for a contrary view see, e.g., Attew, M., 'Teleological Interpretation and Land Law', (1995) 58 MLR 696–701 (considering, *inter alia*, the broad definition of 'supply of goods' and 'supply of services' for the purposes of the VAT legislation). Accordingly, if quota clauses relate to the creation or transfer of interests in land, they will fall outside the ambit of the Unfair Contract Terms Act 1977; and there is a good possibility that they will fall outside the ambit of the Council Directive and national implementing regulations. Moreover, even if the tenant successfully shows that quota clauses do not relate to land or that the Council Directive covers land contracts, the detailed legislative criteria present further hurdles. E.g., the Council Directive would require, *inter alia*, that the landlord is acting for purposes relating to his trade, business, or profession; and that the tenant is not so acting. Besides, it applies only to terms which have not been individually negotiated, in particular those within the context of pre-formulated standard contracts. That having been said, where certain aspects of a term or one or more specific terms have been individually negotiated, it will still apply to the rest of the contract if on an overall assessment it remains a pre-formulated standard contract.

[37] For rent review under the AHA 1986, generally, see, e.g., Muir Watt, J., *Agricultural Holdings*, at 43–64; Densham, H. A. C., *Scammell and Densham's Law of Agricultural Holdings*, at 92–119; and Rodgers, C. P., *Agricultural Law*, at 80–97. For consideration of milk quotas in this context, see, e.g., Wood, D. *et al.*, *Milk Quotas: Law and Practice*, at 31–6 and 59; Densham, H. A. C., *Scammell and Densham's Law of Agricultural Holdings*, at 97–9; and Rodgers, C. P., *Agricultural Law*, at 96–7. There is still doubt as to the extent to which the parties may contract out of s.12 and Sch.2: see, e.g., Densham, H. A. C., *Scammell and Densham's Law of Agricultural Holdings*, at 115–7 (and *1993 Supplement*, at A4–5). However, in the unreported English County Court case of *Trollope* v. *Trustees of Ceawlin's Fund* it was held possible for a landlord and tenant to substitute their own formula without contravening public policy. By contrast, express statutory authority is granted to contract out of the specific provisions applicable to milk quotas on rent review enacted by the Agriculture Act 1986, s.15: see, in particular, s.15(1). For rent review under the ATA 1995, generally, see, e.g., Evans, D., *The Agricultural Tenancies Act 1995*, at 8/9–14; and Sydenham. A. and Mainwaring, N., *Farm Business Tenancies: Agricultural Tenancies Act 1995*, at 45–53.

[38] For useful dicta on this point, see *Broadland Estates Properties Ltd.* v. *Mann*, [1994] SLT (Land Ct.) 7. Presumably any apportionment would be made on the basis of the rules applicable on the transfer of quota. Where quota has been transferred to the tenant wholly or partly at his expense, the need for apportionment in the case of such composite holdings is expressly imposed by statute. The apportionment in such circumstances is to be conducted as on a change of occupation under the milk quota legislation: the Agriculture Act 1986, s.15(1)(b).

'all relevant factors'. These include the terms of the tenancy, the character and situation of the holding, the productive capacity of the holding and its related earning capacity, and the current level of rents for comparable lettings. It would be hard to deny that the presence of milk quota attached[39] to the holding is a 'relevant factor', the expression having been widely interpreted to include such matters as, for example, income from farm shops.[40] Of the criteria expressly set out in the Schedule particular attention has focused on the 'productive capacity of the holding' and 'its related earning capacity'.[41] The productive capacity of the holding would seem to remain unimpaired, notwithstanding the imposition of milk quotas, in that they give rise to a superlevy on over-production rather than prohibit production. However, milk quotas cannot but affect the related earning capacity, as was acknowledged by the Minister of Agriculture during the passage of the AHA 1986 through Parliament.[42] Besides, in *Wachauf* v. *Bundesamt für Ernährung und Forstwirtschaft* the Advocate-General saw the value of quota being translated into higher rental values.[43]

In the case of tenancies governed by the ATA 1995, a statutory rent review machinery is again to apply (notwithstanding any agreement to the contrary), subject to two exceptions. The first exception is where the instrument creating the tenancy expressly states that there will be no rent review during the tenancy. The second is where the instrument creating the tenancy provides that the rent be varied at one or more specified times by or to a specified amount or, alternatively, in accordance with a specified formula which does not preclude a reduction and which does not require or permit the exercise by any person of any judgment or discretion.[44] Accordingly, there is some scope for negotiation; and, on entering into the agreement, the parties would be free to stipulate, for example, that the statutory rent review machinery should not apply and that the rent should be increased in line with prices paid on certain specified days for permanently or temporar-

---

[39] However, the extent to which milk quota 'attaches' to a holding has recently been analysed by the Special Commissioners in *Cottle* v. *Coldicott*, [1995] SpC 40. In particular, quota transferred through the medium of short-term tenancies was held to be a separate asset in the hands of the producer for the purposes of the Capital Gains Tax legislation. See, generally, Ch. 6, 2.2. Nevertheless, it has not been widely suggested that separate sums should be charged by landlords in respect of the quota and the land.

[40] *Enfield L.B.C.* v. *Pott*, [1990] 34 EG 60.

[41] 'Productive capacity' means 'the productive capacity of the holding (taking into account fixed equipment and any other available facilities on the holding) on the assumption that it is in the occupation of a competent tenant practising a system of farming suitable to the holding'; and 'related earning capacity' means 'the extent to which, in the light of that productive capacity, a competent tenant practising such a system of farming could reasonably be expected to profit from farming the holding': the AHA 1986, Sch.2, para.1(2)(a) and (b).

[42] *Hansard* (HC) Vol.61, Col.378.

[43] Case 5/88, [1989] ECR 2609, [1991] 1 CMLR 328.   [44] The ATA 1995, s.9.

ily transferred quota.[45] However, if the statutory machinery does apply, the criteria are similar to those applicable under the AHA 1986. In particular, the arbitrator is to determine 'the rent properly payable'. For these purposes such rent is the amount 'at which the holding might reasonably be expected to be let on the open market by a willing landlord to a willing tenant'. The arbitrator is to take into account 'all relevant factors', which, as seen, would seem wide enough to embrace the presence of milk quota attached to the holding.[46]

Within the umbrella of these provisions milk quota may on occasion fall to be disregarded in the determination of the rent. The clearest disregard is contained in section 15 of the Agriculture Act 1986, which applies in the case of arbitrations under the AHA 1986, subject to any agreement between the parties. Where section 15 does apply, the arbitrator must first ascertain the amount of quota which attaches to the land subject to review, if necessary carrying out an apportionment under the milk quota transfer rules. He must then disregard any increase in the rental value which is attributable to quota transferred to the tenant by a transaction funded wholly or partly at the tenant's expense, with a corresponding proportion of the quota falling out of account where the tenant paid a part rather than all of the cost. The provision also benefits certain successors and assignees of the original tenant, and certain tenants who have sublet.[47] By way of illustration, where the original tenant incurred the costs of the quota transfer, a subsequent assignee of the tenancy may have the right to require that such quota be disregarded on rent review.

As indicated, it is necessary for the arbitrator to ascertain whether the transfer was effected entirely or in part at the tenant's expense, the disregard in the latter case being restricted to a corresponding proportion of the quota transferred. Although this task for the arbitrator is of considerable importance, it may not always be easy to carry out in practice. The statute does provide some clarification, stating that any payment for the grant or assignment of the tenancy, or any previous tenancy, will not qualify.[48]

---

[45] For an early example of a similar formula, see *Bolesworth Estate Co. Ltd.* v. *Cook*, (1996) 116 NLJ 1318. It would also seem open to the parties to agree a formula outside the terms of s.9; but in these circumstances either could opt to follow the statutory procedure rather than the agreed formula. See, e.g., Evans, D., *The Agricultural Tenancies Act 1995*, at 8/12; and Sydenham, A. and Mainwaring, N., *Farm Business Tenancies: Agricultural Tenancies Act 1995*, at 47.

[46] The ATA 1995, s.13(1) and (2). There is, under the ATA 1995, no reference to, *inter alia*, 'related earning capacity'.

[47] The Agriculture Act 1986, s.15(2)(b) and (c).

[48] Ibid., s.15(2)(a). This provision would seem to exclude the payment of a premium on the grant of a tenancy; but the position may be complicated where payment was expressed to be made specifically for quota (e.g., where a new tenant was required to pay to his landlord a sum in respect of quota equal to the amount which the landlord had paid to the departing tenant by way of compensation). On this aspect see Wood, D. *et al.*, *Milk Quotas: Law and Practice*, at 34–5; and Gregory, M. and Sydenham, A., *Essential Law for Landowners and Farmers*, at 121.

However, disputes may arise where the cost of the transaction is paid not by the tenant but by a farming partnership of which he is a member or by a family company of which he is a shareholder and/or director. Not only is the partnership or company more likely to be the registered producer and occupier for the purposes of both the Community and United Kingdom regulations; but it would also be natural for such trading entities to fund the purchase. Accordingly, on a strict interpretation it could be argued that the transaction was not effected at the tenant's expense.[49] Allied to this requirement is the need for a 'transaction', an expression not defined in the legislation. However, it is not clear whether it is necessary to show a legally enforceable contract.[50] The criterion would not appear to be satisfied where a tenant, by changing the areas used for milk production, brings quota on to the tenanted land subject to rent review from other land within his composite Euro-holding.[51]

Apart from the specific provision in section 15 of the Agriculture Act 1986, tenants governed by the AHA 1986 may have difficulty establishing that milk quota should be disregarded on rent review. In particular, little assistance would now seem available from the direction that the arbitrator must not take into account any increase in the rental value of the holding which is attributable to tenant's improvements or fixed equipment (other than improvements executed or equipment provided under an obligation imposed on a tenant by the tenancy).[52] In the case of *Broadland Properties Estates Ltd.* v. *Mann* the argument was advanced that any effect 'allocated' quota may have in increasing the rental value should not be taken into account on the basis that it constituted a tenant's improvement.[53] The Court noted that under the statute transferred quota alone falls to be disregarded, with the clear corollary that regard must be had to allocated quota. Further,

---

[49] The dangers would seem less where the trading entity is a partnership as opposed to a company. In the former case the tenant could at least claim that he had made payment by virtue of his status as a member of the partnership, whereas in the latter case he could be materially prejudiced by the presence of the corporate veil.

[50] See, however, the broad interpretation given 'transaction' in the context of an exceptional hardship claim: *R.* v. *Dairy Produce Quota Tribunal for England and Wales*, ex p. *Lifely*, [1988] 27 EG 79.

[51] See Wood, D. *et al.*, *Milk Quotas: Law and Practice*, at 34–5; and Gregory, M. and Sydenham, A., *Essential Law for Landowners and Farmers*, at 121.

[52] For these purposes 'tenant's improvements' are defined as 'any improvements which have been executed on the holding, in so far as they were executed wholly or partly at the expense of the tenant (whether or not that expense has been or will be reimbursed by a grant out of money provided by Parliament or local government funds) without any equivalent allowance or benefit made or given by the landlord in consideration of their execution'; and 'tenant's fixed equipment' is defined as 'fixed equipment provided by the tenant': the AHA 1986, Sch.2, para.2(2)(a) and (b). Moreover, 'high farming' qualifies as a 'tenant's improvement': ibid., para.2(4). For general discussion, see, e.g., Muir Watt, J., *Agricultural Holdings*, at 54–5; Densham, H. A. C., *Scammell and Densham's Law of Agricultural Holdings*, at 103–7; and Rodgers, C. P., *Agricultural Law*, at 95–6.

[53] [1994] SLT (Land Ct.) 7 (considering the Scottish equivalent of the Agriculture Act 1986, s.15). For 'allocated quota', see the Agriculture Act 1986, Sch.1, para.1(1).

allocated quota could not be a tenant's improvement, in that it could not be said to have been 'executed on the holding' wholly or partly at the tenant's expense. A similar line of reasoning had in fact already been adopted by the County Court in *Marshall* v. *Hughes*, where the tenant failed to establish that milk quota qualifies as a tenant's improvement under the head of 'high farming'.[54] However, the focus of that judgment was directed less to legal principle than failure to satisfy the requirement that there be continuous adoption of such a system of farming.[55]

Under the ATA 1995 it has been seen that the parties may stipulate in the instrument creating the tenancy that there is to be no rent review during the tenancy, or that there is to be variation at a specified time or times by or to a specified amount or in accordance with a specified formula which does not preclude a reduction and which does not require or permit the exercise of any judgment or discretion. In default of such stipulation, the statutory machinery applies[56] — in which case the arbitrator is again to disregard tenant's improvements, subject to certain exceptions.[57] Of some importance in this context is the definition of 'tenant's improvement', which covers not just physical improvements but 'any intangible advantage which — (i) is obtained for the holding by the tenant by his own effort or wholly or partly at his own expense, and (ii) becomes attached to the holding'.[58] This definition is widely apprehended to include milk quotas.[59] Indeed, milk quotas have been been expressly termed an 'advantage' by the European Court;[60] and as recently as 1994 the European Court could refer to 'the

[54] Unreported, Pontypool County Court, 12 July 1992. See Densham, H. A. C., *Scammell and Densham's Law of Agricultural Holdings, 1993 Supplement*, at A10.

[55] See the AHA 1986, Sch.2, para.2(4).

[56] As also seen, either party may still have recourse to the statutory machinery where a rent review formula has been agreed between landlord and tenant formula outside the terms of the ATA 1995, s.9. Should the matter go to arbitration in accordance with the statutory machinery, the procedure applicable is that under the Arbitration Acts 1950–1979 as opposed to that under the AHA 1986.

[57] The exceptions are, first, any improvement provided under an obligation imposed on the tenant by the terms of the tenancy or any previous tenancy and which arose on or before the beginning of the tenancy in question; secondly, any improvement to the extent that allowance or benefit has been made or given by the landlord in consideration of its provision; and, thirdly, any improvement to the extent that the tenant has received compensation for it from the landlord: the 1995 Act, s.13(3).

[58] Ibid., s.13(5) and 15.

[59] See, e.g., Moody, J., *Agricultural Tenancies Act 1995 — a Practical Guide*, at 29; and Sydenham, A. and Mainwaring, N., *Farm Business Tenancies: Agricultural Tenancies Act 1995*, at 67.

[60] See, e.g., Case 44/89, *Von Deetzen II*, [1991] ECR I–5119, [1994] 2 CMLR 487, at I–5156 (ECR) and 511 (CMLR); and Case 2/92, *R.* v. *Ministry of Agriculture, Fisheries and Food*, ex p. *Bostock*, [1994] ECR I–955, [1994] 3 CMLR 547, at I–984 (ECR) and 570 (CMLR). See also the Opinion of the Advocate-General in Case 5/88, *Wachauf* v. *Bundesamt für Ernährung und Forstwirtschaft*, [1989] ECR 2609, [1991] 1 CMLR 328, at 2630 (ECR) and 342 (CMLR) (where he considered milk quota an 'intangible asset'); and the Irish case of *Lawlor* v. *Minister for Agriculture*, [1990] 1 IR 356, [1988] IRLM 400, [1988] 3 CMLR 22, at 374 (IR), 414 (IRLM), and 38 (CMLR) (where Murphy J saw milk quota as 'a valuable intangible asset').

general principle that every reference quantity is to remain attached to the land in respect of which it is allocated'.[61]

Finally, one further aspect may be highlighted. As a general rule, high rents may be obtained from a tenant under the AHA 1986 who attracted a high allocation of milk quota by reason of production in the reference year. The amount of quota will be a relevant factor and, in the absence of transferred quota, the disregard under section 15 of the Agriculture Act 1986 will be of no assistance. However, the situation of such a tenant should be mitigated by the receipt of a greater sum in respect of compensation on the termination of the tenancy. Further, if the tenancy agreement so permits (and this would be another 'relevant factor'), the tenant can derive a substantial income by the temporary transfer of quota to third parties. Indeed, in the light of the high prices paid, a landlord might feel unjustly prejudiced if the rent did not reflect this source of income where available to his tenant. By contrast, if the allocation of quota was lower than average, then the related earning capacity will be impaired and the level of rent obtainable on arbitration reduced. That having been said, the tenant should receive a lesser sum in respect of compensation on termination; and his scope for entering into profitable temporary transfers will likewise be curtailed.

4. COMPENSATION

4.1. General

Soon after the introduction of the milk quota system it was recognized that tenants might suffer both economic and social difficulties as a result of their more limited interest in the land.[62] Most notably, the transfer rules as initially implemented rendered it impossible for them to retain their milk quota on the termination of their tenancy, even if they wished to continue milk production elsewhere. This concern, as indicated, was addressed by the new Article 7(4) of Council Regulation (EEC) 857/84, which granted Member States the option to provide that, where a lease was due to expire without the possibility of extension on similar terms, all or part of the

---

[61] Case 98/91, *Herbrink* v. *Minister van Landbouw, Natuurbeheer en Visserij*, [1994] ECR I–223, [1994] 3 CMLR 645, at I–253 (ECR) and 669–70 (CMLR). However, mention should also be made of *Cottle* v. *Coldicott*, [1995] SpC 40. In that case the Special Commissioners held milk quota to be a separate asset for Capital Gains Tax purposes when transferred through the medium of short-term tenancies (while taking care to confine their decision to its particular context). See, generally, Ch.6, 2.2.

[62] By contrast, for the view that the landlord has no more than an 'expectation' of quota at the end of the lease, see *Re the Küchenhof Farm*, [1990] 2 CMLR 289 (a decision of the German Administrative Court).

corresponding reference quantity be put at the disposal of the departing lessee, if he intended to continue milk production.[63]

Following implementation of Article 7(4), it fell to the European Court in *Wachauf* v. *Bundesamt für Ernährung und Forstwirtschaft* to consider whether the combination of that provision and the available outgoers' schemes was sufficient to ensure protection of the tenant's fundamental rights.[64] The Court stated that, in safeguarding such rights, it was required to look to the constitutional traditions common to Member States and to guidelines supplied by international treaties for the protection of human rights on which Member States had collaborated or to which they were signatories.[65] However, such fundamental rights were not absolute. Rather, they must be considered in relation to their social function. Accordingly, 'restrictions may be imposed on the exercise of those rights, in particular in the context of a common organization of a market, provided that those restrictions in fact correspond to objectives of general interest pursued by the Community and do not constitute, with regard to the aim pursued, a disproportionate and intolerable interference, impairing the very substance of those rights'.[66] Besides, it was emphasized that, when implementing Community rules, Member States must as far as possible apply those rules in accordance with the requirements of the protection of fundamental rights.

Addressing specifically the position of a tenant whose lease had expired, there was acceptance that in principle the reference quantity would return to the landlord where he retook possession. However, if the Community rules had the effect of depriving such a tenant without compensation of 'the fruits of his labour and of his investments in the tenanted holding', then they would be incompatible with the requirements of the protection of fundamental rights.[67] On considering the provisions in detail, the Court believed that the discretion granted to Member States was sufficient to prevent a breach of this kind. In mitigation of the general principle they could choose to make available reference quantities to departing tenants who intended to continue milk production, and they could choose to operate outgoers' schemes. Consequently, the Community legislation question was valid.[68] Nevertheless, consideration was given to the possibility of compensation schemes. Under such schemes a departing tenant might obtain payment in

---

[63] [1984] OJ L90/13, as inserted by Council Reg. (EEC) 590/85, [1985] OJ L68/1. That the provision applied equally to SLOM quota was confirmed in Case 98/91, *Herbrink* v. *Minister van Landbouw, Natuurbeheer en Visserij*, [1994] ECR I–223, [1994] 3 CMLR 645.

[64] Case 5/88, [1989] ECR 2609, [1991] 1 CMLR 328.

[65] See, generally, e.g. Lasok, D., *Law and Institutions of the European Union*, at 160–4. See also, e.g., Case 4/73, *Nold* v. *Commission*, [1974] ECR 491, [1974] 2 CMLR 338; and Case 44/79, *Hauer* v. *Rheinland-Pfalz*, [1979] ECR 3727, [1980] 3 CMLR 42.

[66] Case 5/88, [1989] ECR 2609, [1991] 1 CMLR 328, at 2639 (ECR) and 349 (CMLR).

[67] Ibid.   [68] On this aspect, see Barents, R., *The Agricultural Law of the EC*, at 358.

accordance with his contribution to the building-up of milk production on the holding, although specific provisions were not addressed. As later observed by the Advocate-General in *R.* v. *Minister for Agriculture, Fisheries and Food*, ex parte *Bostock*, 'the Court confined itself to laying down a general principle from which it is not readily possible, in my view, to infer the detailed conditions for when compensation is payable and the principles that are to apply in determining the amount of such compensation'.[69]

It is of note that this aspect of the judgment in *Wachauf* diverged from the Opinion of the Advocate-General. He too adopted the line that on the termination of a tenancy the milk quota would revert to the landlord (or other incoming occupier) under the general principle; but he was not convinced that the existing legislation provided adequate protection for the tenant. Indeed, the permanent loss to the tenant of the use and value of the quota on the termination of a tenancy could be viewed as an expropriatory measure and breach of the principle of the respect for the right to property. In particular, the two hallmarks of expropriatory measures were satisfied: the legislation as it stood would result in the deprivation of all appreciable economic value in the asset; and that deprivation would be permanent.[70] In support of this view he emphasized that Article 7(4) of Council Regulation (EEC) 857/84 was optional and that, in any event, it was of no assistance to departing farmers who preferred to give up milk production (for example, with a view to retirement or pursuing a different occupation).[71] In the course of expressing reservations that the agricultural holdings legislation of Member States could be relied upon to redress the balance in favour of the tenant, he drew attention to the fact that the United Kingdom had felt it appropriate to enact specific measures (in the Agriculture Act 1986) so as to introduce a compensation scheme. Despite this difference in approach, the Advocate-General was equally reluctant to spell out the detailed framework of any such compensation scheme. However, he could conclude as follows:

The principle of respect for the right to property guaranteed by the Community legal order requires Member States to provide for financial compensation by the landlord to a tenant farmer who, on expiry of the lease of a holding, loses the right to exploit the quota, in a case where, having regard to the particular situation of the tenant farmer, failure to provide for compensation would result in a breach of that principle.[72]

---

[69] Case 2/92, ECR I–955, [1994] 3 CMLR 547, at I–967 (ECR) and 558 (CMLR).

[70] See also, e.g., Case 44/79, *Hauer* v. *Rheinland-Pfalz*, [1979] ECR 3727, [1980] 3 CMLR 42. For these purposes his Opinion provides full recognition of the value of quota.

[71] See also Case 2/92, *R.* v. *Minister for Agriculture, Fisheries and Food*, ex p. *Bostock*, [1994] ECR I–955, [1994] 3 CMLR 547.

[72] Case 5/88, [1989] ECR 2609, [1991] 1 CMLR 328, at 2632 (ECR) and 344–5 (CMLR). See also the Opinion of the Advocate-General in Case 121/90, *Posthumus* v. *Oosterwoud*, [1991] ECR I–5833, [1992] 2 CMLR 336; and the Opinion of the Advocate-General in Case 177/90, *Kühn* v. *Landwirtschaftskammer Weser-Ems*, [1992] ECR I–35, [1992] 2 CMLR 242.

None the less, the European Court in *R. v. Ministry of Agriculture, Fisheries and Food*, ex parte *Bostock* had no hesitation in following its judgment in *Wachauf* v. *Bundesamt für Ernährung und Forstwirtschaft*. In express terms it stated that neither the Community regulations nor the general principles of Community law imposed on Member States a requirement to introduce a compensation scheme for departing tenants; and, besides, the same regulations and general principles conferred on tenants no directly enforceable right as against their landlords.[73] These points would, accordingly, seem settled beyond doubt.

As indicated, the United Kingdom did not implement Article 7(4) of Council Regulation (EEC) 857/84, with the result that on the termination of a tenancy the general transfer rules continued to apply, the milk quota registered in the name of the tenant passing to the incoming occupier (in all probability the landlord or an incoming tenant).[74] However, as also indicated, to alleviate the effect of this general rule a scheme of compensation was implemented under the Agriculture Act 1986, this solution attracting the attention of the Advocate-General in *Wachauf* v. *Bundesamt für Ernährung und Forstwirtschaft*. Moreover, the parties in *R. v. Minister for Agriculture, Food and Fisheries*, ex parte *Bostock* accepted that the Agriculture Act 1986 complied with Community rules. The dispute concerned rather the position of tenants whose tenancies expired before it came into force.[75] In addition, following replacement of Council Regulation (EEC) 857/84 by Council Regulation (EEC) 3950/92 the provision of such a compensation scheme met well the requirement that, where there is no agreement between the parties, on the termination of non-renewable leases or in situations with comparable legal effects, reference quantities are to be transferred in whole or in part to the producers taking over the holdings, 'in accordance with provisions adopted or to be adopted by the Member States taking account of the legitimate interests of the parties'.[76]

Before considering the detailed provisions of the Agriculture Act 1986, it may be noted that the eligibility criteria are closely circumscribed. In particular, no compensation is payable in respect of tenancies expiring before the Act came into force on 25 September 1986, and the Act does not apply to farm business tenancies under the ATA 1995. The tenant seeking to qualify for statutory compensation under the Agriculture Act 1986 is thus faced with material constraints; and failure to meet any of

---

[73] Case 2/92, [1994] ECR I–955, [1994] 3 CMLR 547.

[74] [1984] OJ L90/13, as inserted by Council Reg. (EEC) 590/85, [1985] OJ L68/1.

[75] Case 2/92, [1994] ECR I–955, [1994] 3 CMLR 547.

[76] [1992] OJ L405/1, Art.7(2). For comment on the effects of the new provision, see the Opinion of the Advocate-General in Case 2/92, *R. v. Ministry of Agriculture, Fisheries and Food*, ex p. *Bostock*, [1994] ECR I–955, [1994] 3 CMLR 547. In particular, he stated that Art.7(2) 'may signify that the Council considered that it was necessary to lay down expressly such an obligation for the Member States but it may also merely signify that the Council considered it appropriate to lay down that duty in express terms, even though it was already implicit in the general principles of Community law': at I–964 (ECR) and (CMLR).

the eligibility criteria leaves compensation for tenants under the AHA 1986 dependent on agreement with their landlords.

4.2. The Agriculture Act 1986

4.2.1. General

The Agriculture Act 1986 has the objective of ensuring that on the termination of their tenancies certain qualifying tenants are not deprived without compensation of the fruits of their labours and investments in the tenanted land.[77] To achieve this objective, compensation may be payable under one or more of three heads. First, if the tenant has secured an allocation of milk quota in respect of the tenanted land higher than the 'standard quota' as determined under the Agriculture Act 1986 and requisite regulations, then he may receive compensation in respect of that 'excess over standard quota'. Secondly, he may receive compensation for a proportion of the standard quota (or of the quota actually allocated, if equal to or less than the standard quota) to reflect his provision of dairy improvements and/or fixed equipment — the 'tenant's fraction'. Both these claims may, broadly, be said to relate to 'allocated quota'. Thirdly, he may claim for 'transferred quota' where he has paid all or part of the cost of a transaction by which quota has been transferred to tenanted land. If he has paid only part of the cost, compensation is payable to the extent of that contribution. Having established the general eligibilty criteria, each of these claims will be considered in turn.

4.2.2. Eligibility criteria

As a preliminary question, it is necessary to determine whether the tenant satisfies the various eligibility criteria, which are both complex and impose considerable restrictions. First, only certain forms of tenancy carry a right to compensation.[78] Those which meet this first requirement are certain forms of tenancy under the AHA 1986, namely: tenancies from year to year; fixed-term tenancies for two years or more which continue from year to year under section 3 of the AHA 1986; arrangements which would have effect as

---

[77] S.13 and Sch.1. See, generally, Wood, D. *et al.*, *Milk Quotas: Law and Practice*, at 40–8 and (for a calculation) 60–2; Wood, D. *et al.*, *The Handbook of Milk Quota Compensation*, *passim*; Muir Watt, J., *Agricultural Holdings*, at 225–42; Densham, H. A. C., *Scammell and Densham's Law of Agricultural Holdings*, at 318–35; Gregory, M. and Sydenham, A., *Essential Law for Landowners and Farmers*, at 121–6; and Rodgers, C. P., *Agricultural Law*, at 325–36. For Parliamentary debate, see, in particular, *Hansard* (HC) Vol.95, Cols.1036–1103.

[78] See the definition of 'tenancy': the Agriculture Act 1986, Sch.1, para.18(1).

tenancies from year to year under section 2 of the AHA 1986, were it not for prior approval by the Minister;[79] and tenancies which would continue from year to year under section 3 of the AHA 1986, were it not for prior approval by the Minister under section 5 of the same Act.[80] Accordingly, there is no right to compensation in the case of *Gladstone* v. *Bower* agreements or grazing and/or mowing agreements for a specified period of the year. Moreover, it is expressly provided by the ATA 1995 that the statutory compensation scheme under the Agriculture Act 1986 does not apply in the case of farm business tenancies.[81] Since the farm business tenancy régime is that applicable where the tenancy begins on or after 1 September 1995 (subject to certain limited exceptions), the tenant taking a tenancy beginning on or after that date must now seek to secure appropriate terms for his protection in the agreement itself rather than rely on any statutory scheme.[82]

Secondly, the tenancy must have terminated and the tenant must have quitted the agricultural holding.[83] However, it is now established that no claim for compensation can be made where termination took place before the Agriculture Act 1986 came into force on 25 September 1986.[84] Indeed, this was the specific question before the European Court in *R.* v. *Ministry of Agriculture, Fisheries and Food*, ex parte *Bostock*, the applicant having surrendered his holding on 25 March 1985.[85] As seen, the judgment confirmed that neither the Community regulations nor the general principles of Community law required a Member State to introduce a scheme of compensation for departing tenants. Likewise, they conferred no direct right to such compensation on the tenants themselves. In particular, the Court dismissed argument that, in the absence of such compensation, tenants who quitted before 25 September 1986 would be the victims of discrimination as compared with those who quitted on or after that date. The principle

---

[79] It may be noted that the definition employs the word 'arrangement' rather than 'tenancy'; and it has been suggested that this is with the intention of including certain licences: Wood. D. *et al.*, *The Handbook of Milk Quota Compensation*, at 7.

[80] In addition there are specific provisions to cover succession tenancies, tenancies which have been assigned, and the circumstances of tenants who have granted subtenancies.

[81] The ATA 1995, s.16(3). For compensation under the ATA 1995, see Ch.5, 4.3.

[82] For the exceptions, see the ATA 1995, s.4. Those most likely to be relevant in this context are variations which effect a surrender and regrant of existing lettings under the AHA 1986 (s.4(1)(f)); and certain succession tenancies (s.4(1)(b),(c), and (d)).

[83] 'Termination' for these purposes means 'the cesser of the letting of the land in question or the agreement for letting the land, by reason of effluxion of time or from any other cause': the Agriculture Act 1986, Sch.1, para.18(1). Thus, a surrender of the tenancy would be included.

[84] The Agriculture Act (Commencement) (No.1) Order 1986, S.I.1986 No.1484.

[85] Case 2/92, ECR I–955, [1994] 3 CMLR 547. It may be noted that the applicant did not as at that date have the option of obtaining compensation under an outgoers' scheme. The litigation was also of some importance in the context of judicial review, the Court of Appeal having cause to consider whether or not to extend the application period under RSC Ord.53, r.4(1): [1991] 1 CMLR 681 (High Court); [1991] 1 CMLR 687, [1991] 41 EG 129 (CA).

of equal treatment was held incapable of imposing such retroactive modification of the landlord–tenant relationship.[86]

Thirdly, the tenant must have quota 'registered as his' in relation to a holding consisting of or including the land comprised in the tenancy.[87] This requirement was soon identified as one liable to give rise to practical difficulties.[88] No such difficulties should arise in the simple case where the tenant and registered producer are one and the same (as is likely, for example, where the tenant is a sole trader). By contrast, where the tenant is an individual and the registered producer is a partnership or company, it is less clear that the requirement is met.[89] None the less, there is a strong argument that, if the tenant is a member of such a partnership, it would be for him to make the claim and distribute the compensation in accordance with any terms to that effect in the partnership agreement.[90] If the agreement was drafted after 25 September 1986, in all probability there will be an express provision governing the destination of the compensation. Failing that, the partnership agreement may contain provisions such as those found in *Faulks* v. *Faulks*, where the tenancy was held on trust for the partnership during the continuance of the partnership; and, as decided in *Faulks* v. *Faulks*, entitlement to compensation would presumably pass to the landowning partner.[91] In this event it might be possible to claim that on dissolution allowance should be made in the accounts for any money which the other partner or partners had expended on maintaining or enhancing the value of the tenancy, the claim being based on the principle in *Pawsey* v.

---

[86] It would be hard to deny that the Agriculture Act 1986 imposed on landlords an obligation for the future to pay where there had not been one before; but in *Bostock* the Court appeared reluctant to extend the burden of compensation to landlords whose tenants had departed a decade or so prior to the judgment. The number of potential claimants and the difficulties in collating data may have been factors behind the decision.

[87] Agriculture Act 1986, Sch.1, para.1(1). In this context, as for the purposes of rent review, 'holding' has the same meaning as that ascribed to it by the Community regulations and may therefore extend beyond the land comprised in the tenancy which is being terminated. Should such be the case, apportionment is required to determine the 'relevant quota' attaching to that tenanted land (see Ch.5, 4.2.3.).

[88] See, e.g., Wood, D. *et al.*, *Milk Quotas: Law and Practice*, at 41; and Wood, D. *et al.*, *The Handbook of Milk Quota Compensation*, at 9–10.

[89] Although as a matter of law a partnership does not enjoy separate legal identity, it would seem that the milk marketing boards were prepared to register quota in the trading name of a partnership. Registration in this form has apparently remained acceptable to the Intervention Board since they took over administrative responsibility on 1 Apr. 1994; and, indeed, it may be noted that the definition of 'producer' under the Community legislation refers to 'a natural or legal person or a group of natural or legal persons farming a holding': Council Reg. (EEC) 3950/92, [1992] OJ L405/1, Art.9(c).

[90] See Densham, H. A. C., *Scammell and Densham's Law of Agricultural Holdings*, at 320.

[91] [1992] 15 EG 82. In the tax case of *Cottle* v. *Coldicott* the Special Commissioners expressly stated that in their view *Faulks* v. *Faulks* was rightly decided on its facts, while holding that milk quota transferred through the medium of short-term tenancies constitutes a separate asset for Capital Gains Tax purposes: [1995] SpC 40.

Armstrong[92] and *Miles* v. *Clarke*.[93] Although in *Faulks* v. *Faulks* Chadwick J did not feel the evidence merited any such allowance, he found it easy to accept that the principle could apply; and he cited, among other examples, circumstances where the dairy enterprise had been commenced by the partnership.[94]

If the tenant is an individual and the registered producer is a company, there is a greater risk that the right to compensation may be lost. Although he may also be a director and/or shareholder, the corporate veil would be a significant barrier to any argument that the tenant had milk quota 'registered as his'. In this context an alternative basis of claim may be available. The licence under which the trading company farmed the land could, in the correct circumstances, have been converted into a tenancy from year to year by section 2 of the AHA 1986. The company would then hold as subtenant and could make a claim in its own right; and the head-tenant could in turn recover any amount payable to the company from his landlord.[95]

Fourthly, in the case of compensation for 'allocated quota', the tenant must have had milk quota allocated to him in relation to the land comprised in the holding.[96] For these purposes 'allocated quota' includes not only allocations based upon production during the requisite reference year,[97] but also 'secondary allocations' such as development awards or exceptional hardship awards.[98] In the case of compensation for 'transferred quota', the tenant must either have had milk quota so allocated to him or have been in occupation of the land as tenant on 2 April 1984 (whether or not under the tenancy which is terminating).[99] The wider ambit of this latter provision would include, for example, a person who was tenant as at 2 April 1984 but only transferred milk quota onto the tenanted land and commenced milk production after that date. However, as a general rule a person who took up occupation as tenant after 2 April 1984 would not receive statutory entitlement to compensation even if he transferred quota

---

[92] (1881) 18 Ch.D. 698.   [93] [1953] 1 WLR 537.

[94] One possible solution would be for the name of the tenant to be entered on the register, the tenant then holding the quota on trust for the partners. This arguably satisfies the requirement that the tenant has quota 'registered as his', while at the same time protecting the interests of all partners.

[95] The Agriculture Act 1986, Sch.1, para.4. This may be a dangerous argument to adopt, there being a good chance that the subtenancy will permit the landlord to serve notice to quit based on the breach of any covenant against subletting in the head-tenancy agreement: see, e.g., *Snell* v. *Snell*, (1964) 191 EG 361. However, a more benign view was taken in *Pennell* v. *Payne*, [1995] QB 192, [1995] 2 All ER 592, [1995] 2 WLR 261, [1995] 06 EG 152.

[96] The Agriculture Act 1986, Sch.1, para.1(1)(a).

[97] As already seen, under the general rule wholesale quota was allocated by reference to production in the 1983 calendar year; and direct sales quota by reference to production in the 1981 calendar year. In either case producers affected by exceptional circumstances could make a 'base year revision claim'.

[98] The Agriculture Act 1986, Sch.1, para.6(3) and (4).   [99] Ibid., para.1(1)(b).

onto the tenanted land at his expense. Any payment would be a matter for negotiation with his landlord.

Over and above these general rules, paragraphs 2 to 4 of Schedule 1 to the Agriculture Act 1986 extend the right to compensation beyond tenants who were themselves allocated quota or who were themselves in occupation of the land as tenant on 2 April 1984. The persons who may benefit from these provisions are: first, certain successors on death or retirement; secondly, certain assignees; and, thirdly, certain head-tenants, who may effectively recover from their head-landlord an amount equal to the compensation which they pay to their subtenants.

On certain forms of statutory succession to a tenancy, any entitlement to compensation is passed on to the new tenant.[100] Neither the estate of the deceased tenant nor the retiring tenant receives payment, the successor instead being treated as if the allocation or transfer of milk quota in respect of the tenanted land had been made to him rather than his predecessor.[101] In addition, in the case of claims for transferred quota, the successor is treated as having paid the same amount of the cost of the transfer as his predecessor;[102] and as if he had been in occupation on 2 April 1984 where the former tenant was in occupation on that date.[103] Accordingly, the claim for compensation is postponed until the termination of the succession tenancy (or the second succession tenancy, if such arises). However, the eventual claimant effectively acquires the right to compensation accruing from allocations or transfers prior to that date. By way of example, it is possible for a second successor to enjoy a claim in respect of quota allocated to the original tenant and in respect of quota transferred to the first successor and to himself.

Similar provisions apply in the case of assignees. Where a tenancy has been assigned after 2 April 1984, whether by operation of law or by deed,

---

[100] Ibid., para.2. The forms of statutory succession which trigger these consequences are set out in para.2(1)(a), (b), and (c) — i.e., (1) where, following the death of the previous tenant, the new tenancy was obtained by Agricultural Land Tribunal direction under s.39 of the AHA 1986; (2) where, following the retirement of the previous tenant, the new tenancy was obtained by Agricultural Land Tribunal direction under s.53 of the AHA 1986; (3) where, following a direction under s.39 of the AHA 1986, the new tenancy was granted by agreement to a person entitled under such direction (the AHA 1986, s.45(6)); and (4) where the new tenancy was granted by agreement to a close relative of the deceased tenant (the AHA 1986, s.39(1)(b) or (2)). There are also provisions to cover such circumstances occurring before the AHA 1986 came into force: para.2(3). For a detailed analysis of the succession provisions, generally, see, e.g., Muir Watt, J., *Agricultural Holdings*, at 148–87; Densham, H. A. C., *Scammell and Densham's Law of Agricultural Holdings*, at 217–80; and Rodgers, C. P., *Agricultural Law*, at 151–78.

[101] The Agriculture Act 1986, Sch.1, para.2(2)(a). Also included are any allocations or transfers of quota which are treated as having been allocated or transferred to the former tenant.

[102] Ibid., para.2(2)(b)(i). This extends to sums which the former tenant is treated as paying.

[103] Ibid., para.2(2)(b)(ii). This extends to circumstances where the former tenant is treated as having been in occupation on that date.

any milk quota allocated or transferred to the assignor in respect of the tenanted land is instead treated as if it had been allocated or transferred to the assignee. In the case of transferred quota, the assignee is treated as having paid the same amount of its cost as the assignor; and as if he had been in occupation on 2 April 1984 where the assignor was in occupation on that date.[104] Accordingly, compensation is payable on the termination rather than the assignment of the tenancy.

In the case of subtenancies, Schedule 1 to the Agriculture Act 1986 confers full entitlement to compensation upon subtenants (tenancy being sufficiently broadly defined as to include subtenancies).[105] However, in the absence of the express provision in paragraph 4 of the Schedule, the head-tenant against whom such a claim must be directed could be severely prejudiced. It would not be possible for a head-tenant who had been neither allocated quota in respect of the tenanted land nor in occupation on 2 April 1984 to sustain a claim against his landlord. As a result, in such circumstances he would be obliged to compensate the subtenant without reimbursement on the termination of his own tenancy. Paragraph 4 remedies any potential hardship by providing that the head-tenant may effectively recover from his landlord the amount of compensation paid to the subtenant. This is achieved by treating any milk quota allocated or transferred to the subtenant in respect of the tenanted land as instead having been allocated or transferred to the head-tenant.[106] In the case of claims for transferred quota, the head-tenant is treated as if he had paid so much of the cost as the subtenant; and as if he had been in occupation on 2 April 1984 where the subtenant was in occupation on that date.[107] The statute further addresses the point at which compensation may be recovered from the head-landlord. If the head-tenant himself moves into occupation on the termination of the subtenancy, he will be able to make his claim on the termination of his tenancy in accordance with the general rules. However, if he does not occupy the land on the termination of the subtenancy and departure of the subtenant, then he is taken to have quitted the land at the same time as the subtenant.[108]

---

[104] Ibid., para.3. Again the provisions extend to cover quota treated as having been allocated or transferred to the assignor; and, in the case of transferred quota, to sums treated as having been paid by the assignor and to circumstances where the assignor is treated as having been in occupation on 2 Apr. 1984.

[105] Ibid., para.18(1).

[106] As with successors and assignees, this includes quota treated as having been allocated or transferred to the subtenant.

[107] Once more the provisions cover sums treated as having been paid by the subtenant and circumstances where the subtenant is treated as having been in occupation on 2 Apr. 1984.

[108] Presumably if the head-tenant relets the land without moving into occupation, then the head-landlord must pay compensation without himself being able to move into occupation or charge the incoming tenant for the benefit of the quota: see Densham. H. A. C., *Scammell and Densham's Law of Agricultural Holdings*, at 331.

4.2.3. *Relevant quota*

Before assessing the amount of compensation payable, it is not only necessary to establish eligibility. In addition, the parties must agree the 'relevant quota' which attracts compensation — or, in default of agreement, the question must be determined by an arbitrator. The 'relevant quota' is the amount of the milk quota registered in the name of the tenant which is referable to the land comprised in the tenancy. The need to identify this relevant quota arises from the fact, already observed, that the quota is registered in respect of the producer's Euro-holding, which may embrace a multiplicity of production units, whereas compensation is payable only in respect of the tenanted land. Where the tenanted land is co-extensive with the Euro-holding as defined in the Community regulations, no need for apportionment arises, the relevant quota in such circumstances being 'the milk quota registered in relation to the holding'. However, in other cases the relevant quota is such part of that milk quota as falls to be apportioned to the land subject to the tenancy on the termination of the tenancy.[109] Although Schedule 1 does not provide any guidance as to the manner in which apportionment should be effected, it is assumed that this will be in accordance with the areas used for milk production.[110] Indeed, since the termination of the tenancy would give rise to a change of occupation, it would be necessary to carry out an apportionment for transfer purposes in any event.[111] Following such determination, the tenant is entitled to compensation for allocated quota 'in respect of so much of the relevant quota as consists of allocated quota'; and for transferred quota 'in respect of so much of the relevant quota as consists of transferred quota transferred to him by virtue of a transaction the cost of which was borne wholly or partly by him'.[112]

---

[109] The Agriculture Act 1986, Sch.1, para.1(2). Where only part of the tenanted land is vacated, see para.13.

[110] See, e.g., Wood, D. *et al.*, *Milk Quotas: Law and Practice*, at 41–2; Wood, D. *et al.*, *The Handbook of Milk Quota Compensation*, at 11; Muir Watt, J., *Agricultural Holdings*, at 231; Densham, H. A. C., *Scammell and Densham's Law of Agricultural Holdings*, at 321–2; and Rodgers, C. P., *Agricultural Law*, at 334. It may be noted that any apportionment required for the purposes of rent review is to be conducted on this basis: the Agriculture Act 1986, s.15.(1)(b). For apportionment in accordance with the areas used for milk production, generally, see Ch.4, 6.

[111] Indeed, arbitration may otherwise be required on one form of apportionment for transfer purposes and on another form of apportionment for compensation purposes: Wood, D. *et al.*, *The Handbook of Milk Quota Compensation*, at 11.

[112] The Agriculture Act 1986, Sch.1, para.1(1)(a) and (b).

4.2.4. The heads of claim

4.2.4.1. Outline

The calculation of payment enjoys a complexity in line with its objective of securing for the tenant compensation commensurate with his labours and investments in respect of the tenanted land. As indicated, there are three potential heads of claim, two of these relating broadly to 'allocated quota', namely 'excess over standard quota' and the 'tenant's fraction', and the third relating to 'transferred quota'.[113]

4.2.4.2. Excess over standard quota

Payment for 'excess over standard quota' ensures that the tenant is rewarded for attracting the allocation of an unrepresentatively high number of litres to the tenanted land. While the landlord may for the moment suffer the burden of payment to the departing tenant, he will none the less enjoy the opportunity to exploit that unrepresentatively high number of litres himself. Alternatively, he may seek to recoup the amount of compensation from any incoming tenant to whom the quota is made available or, foregoing such recoupment, charge a higher rent for making the quota available throughout the term of the new tenancy.

Having established the 'relevant quota' referable to the tenanted land, it becomes necessary to determine the standard quota for that land. The amount of standard quota is calculated in accordance with the terms of the Agriculture Act 1986 and the requisite statutory instruments.[114] The general rule is that the standard quota is calculated by multiplying the 'relevant number of hectares' by 'the prescribed quota per hectare'.

Turning first to 'the relevant number of hectares', the expression is defined as:

the average number of hectares of the land in question used during the relevant period for the feeding of dairy cows kept on the land or, if different, the average number of hectares of the land which could reasonably be expected to have been so used (having regard to the number of grazing animals other than dairy cows kept on the land during that period).[115]

---

[113] The Agriculture Act 1986, Sch.1, para.5.

[114] The authority for the making of such statutory instruments is provided by ibid., para.6(7). That first enacted was the Milk Quota (Calculation of Standard Quota) Order 1986, S.I. 1986 No.1530; and amendments have subsequently been effected by the Milk Quota (Calculation of Standard Quota) (Amendment) Order 1987, S.I.1987 No.626, the Milk Quota (Calculation of Standard Quota) (Amendment) Order 1988, S.I.1988 No.653, the Milk Quota (Calculation of Standard Quota) (Amendment) Order 1990, S.I.1990 No.48, the Milk Quota (Calculation of Standard Quota) (Amendment) Order 1991, S.I.1991 No.1994, and the Milk Quota (Calculation of Standard Quota) (Amendment) Order 1992, S.I.1992 No.1225.

[115] The Agriculture Act 1986, Sch.1, para.6(1)(a).

This definition is itself supported by further definitions. For the purposes of the Schedule 'the relevant period' is:

(a) the period in relation to which the allocated quota was determined; or (b) where it was determined in relation to more than one period, the period in relation to which the majority was determined or, if equal amounts were determined in relation to different periods, the later of those periods.[116]

Accordingly, for most wholesale producers the relevant period will be the 1983 calendar year; and for most direct sales producers it will be the 1981 calendar year.[117] Further, in this context 'land used for the feeding of dairy cows kept on the land' specifically excludes 'land used for growing cereal crops for feeding to dairy cows in the form of loose grain'; and 'dairy cows' are specifically defined as 'cows kept for milk production (other than uncalfed heifers)'.[118]

Certain aspects of these provisions may be highlighted. First, they lay down only the general rule, the important exception to which will be considered later. Secondly, the relevant number of hectares is determined upon a different basis to the 'areas used for milk production' as interpreted in *Puncknowle Farms Ltd.* v. *Kane*.[119] For example, land used for growing cereal crops to be fed to the cows in the form of loose grain is excluded for the purposes of calculating the relevant number of hectares, yet would seem to qualify as an 'area used for milk production'.[120] Likewise, the only cows which qualify as dairy cows are those kept for milk production 'other than uncalfed heifers', whereas land for the keeping of all dairy replacements would constitute an 'area used for milk production'. However, the relevant number of hectares clearly does extend beyond the land used by the milking herd on a day-to-day basis so as to include land used for producing feed (but not all forms of feed). Thirdly, no guidance is given as to the method of determining the 'average number of hectares' (in particular, as to whether it should be determined upon a time-apportioned basis or some other

---

[116] Ibid., para.8.

[117] However, by reason of a base year revision claim a wholesale producer's reference year could alternatively be 1981 or 1982; and a direct sales producer's reference year could alternatively be 1982 or 1983.

[118] The Agriculture Act 1986, Sch.1, para.6(5).

[119] [1985] 3 All ER 790, (1985) 275 EG 1283, [1986] 1 CMLR 27. See also the judgments of the European Court in Case 121/90, *Posthumus* v. *Oosterwoud*, [1991] ECR I–5833, [1992] 2 CMLR 336; and Case 79/91, *Knüfer* v. *Buchmann*, [1992] ECR I–6895, [1993] 1 CMLR 692.

[120] As has been noted by commentators, dairy cows are not as a rule fed loose grain: see, e.g., Wood, D. *et al.*, *The Handbook of Milk Quota Compensation*, at 18–19; and Densham, H. A. C., *Scammell and Densham's Law of Agricultural Holdings*, at 324. However, see *Grounds* v. *A-G of the Duchy of Lancaster*, [1989] 21 EG 73 (where both parties and the Court were prepared to accept that such could only be reference to grain grown on the land and then processed before being fed to the dairy cows).

basis).[121] Fourthly, there is again no guidance as to whether use during the relevant period must be exclusive or non-exclusive. If non-exclusive use should qualify, then the tenant is unlikely to receive heavy compensation for excess over standard quota, since the standard quota is calculated by reference to a prescribed quota per hectare which would seem to envisage exclusive use.[122]

The 'prescribed quota per hectare', by which the relevant number of hectares must be multiplied, is defined as 'such number of litres as the Minister may from time to time by order prescribe'.[123] The statutory instruments enacted for this purpose have laid down figures per hectare dependent upon the nature of the land and the breed of cattle. For example, in the case of Channel Island breeds the prescribed quota per hectare is less than for Friesians and Holsteins. However, it is unclear whether the breed in question is that suited to the land or that in fact kept by the tenant.

Once the multiplication has been effected, the tenant will be entitled to payment for the value of any of the allocated quota in excess of the standard quota for the land.[124] If the figures are equal, or the allocated quota is less than the standard quota, then no compensation is payable under this head.[125]

The exception to this general rule is set out in paragraph 6(2) of Schedule 1. The exception applies if 'by virtue of the quality of the land in question or climatic conditions in the area the amount of milk which could reasonably be expected to have been produced from one hectare of the land during the relevant period . . . is greater or less than the prescribed average yield per hectare'. This amount is referred to as the 'reasonable amount'. In such circumstances the calculation of the standard quota is instead effected by multiplying the relevant number of hectares 'by such proportion of the prescribed quota per hectare as the reasonable amount bears to the prescribed average yield per hectare'.[126] As with the 'prescribed quota per hectare', the 'prescribed average yield per hectare' is fixed by statutory instrument; and in practice the Minister has used the same statutory instruments to fix both figures (those for the prescribed average yield per hectare being significantly higher). That having been said, the two terms are to be clearly distinguished. Accordingly, by way of example, it will be in the interests of a landlord to argue that the alternative method is applicable on the basis that the reasonable amount is *greater* than the prescribed average yield per hectare. This would increase the standard quota and

[121] See, e.g., Densham, H. A. C., *Scammell and Densham's Law of Agricultural Holdings*, at 325; and *1993 Supplement* at A18–19.
[122] See, e.g., Wood, D. *et al.*, *The Handbook of Milk Quota Compensation*, at 19.
[123] The Agriculture Act 1986, Sch.1, para.6(1)(b).   [124] Ibid., para.5(2)(a)(ii).
[125] Ibid., para.5(2)(b) and (c).
[126] For these purposes the 'relevant period' and the 'prescribed quota per hectare' have exactly the same meaning as for calculation under the general rule.

reduce the excess over standard quota in respect of which he will be liable for compensation.[127]

The circumstances in which this alternative method may be brought into play were examined by the County Court in *Surrey County Council* v. *Main*.[128] The judge emphasized that the difference between the 'reasonable amount' and the 'prescribed average yield per hectare' must be caused either by the quality of the land in question or by climatic conditions in the area (or both); and that, as a result, there is no need for an arbitrator to ascertain the reasonable amount if satisfied 'on the evidence without more, that the quality of the land and the climatic conditions are not so unorthodox that they would affect the milk yield either way'.[129] In this conclusion he was reinforced, *inter alia*, by the fact that, unless due weight was given to such limitation, the alternative calculation under paragraph 6(2) would as good as render redundant the orthodox calculation under paragraph 6(1): the only other condition required to trigger the operation of paragraph 6(2) was that the reasonable amount was greater or less than the prescribed average yield per hectare — a condition which would almost always be satisfied.

The provisions governing the assessment of the reasonable amount have also fallen to be considered by the Court of Appeal in the case of *Grounds* v. *Attorney-General of the Duchy of Lancaster*.[130] Central to the dispute was whether the arbitrator should disregard concentrates when assessing the amount of milk which could reasonably be expected to have been produced from one hectare of the land during the relevant period. In particular, with a view to reducing the reasonable amount (and thereby increasing the excess over standard quota), the tenant argued that when making such calculation it was to be assumed that the cattle were fed only grass or hay produced on the land in question — and perhaps bought-in hay. The Court of Appeal held that as a matter of law the arbitrator was entitled to use his own knowledge and expertise to decide the course of action which a reasonably skilful and successful farmer would have undertaken during the relevant period; and, if it was his conclusion that such a farmer would normally feed concentrates, the effect of concentrates could be taken into account for the purposes of calculating the reasonable amount.

Finally, before the standard quota can be compared with the amount of allocated quota (for the purposes of ascertaining the number of litres, if any, in respect of which the tenant may be entitled to compensation), some adjustment of the figures may be required. This need arises where the

---

[127] The landlord so argued in, e.g., *Surrey County Council* v. *Main*, [1992] 06 EG 159.

[128] Ibid.

[129] Ibid., at 160. However, the arbitrator may decide to do the calculation, see whether there is any material difference, and then 'look to see if the difference is due, wholly or in part, to the quality of the land or the climatic conditions'.

[130] [1989] 21 EG 73.

relevant quota includes milk quota allocated following an award by a Dairy Produce Quota Tribunal, but the award has not been allocated in full.[131] In these circumstances the standard quota for the land is reduced by an amount equal to the shortfall (the reduction being carried out on a proportionate basis where only part of the quota allocated following the award is relevant quota).

4.2.4.3. Tenant's fraction

The objective of the claim for the 'tenant's fraction' is to reward the tenant for any contribution which he has made through the provision of dairy improvements and/or fixed equipment.[132] Again the detailed rules are complex; but, in effect, the extent of the tenant's contribution is recognized by his receiving a proportionate share, or, as the legislation terms it, the 'tenant's fraction', of either the standard quota or the allocated quota. Whether this tenant's fraction is applied to the standard quota or the allocated quota is dependent upon the respective amount of the standard or allocated quota. If the allocated quota exceeds the standard quota for the land, the tenant's fraction is applied to the standard quota. However, if the allocated quota is equal to or less than the standard quota, then the tenant's fraction is applied to the allocated quota; and, if the allocated quota is less than the standard quota, the tenant's entitlement is scaled down in such proportion as the allocated quota bears to the standard quota.[133] Accordingly, where the allocated quota is less than the standard quota, the tenant is doubly penalized: not only is he excluded from any claim for excess over standard quota, but any claim in respect of the tenant's fraction is reduced commensurately.

The calculation of the tenant's fraction is made in accordance with paragraph 7 of Schedule 1. The numerator of the fraction is 'the annual rental value at the end of the relevant period of the tenant's dairy improvements and fixed equipment'.[134] The denominator is 'the sum of that value and such part of the rent payable by the tenant in respect of the relevant period as is attributable to the land used in that period for the feeding, accommodation or milking of dairy cows kept on the land'.[135] As with the calculation of excess over standard quota, supplementary provisions expand upon and clarify the basic definitions. First, the rental value of the tenant's dairy

---

[131] The Agriculture Act 1986, Sch.1, para.6(3). For these purposes milk quota allocated following an award includes quota allocated for the reason that the amount awarded had not originally been awarded in full: ibid., para.6(4).

[132] As stated by Glidewell LJ in *Grounds* v. *A-G of the Duchy of Lancaster*, [1989] 21 EG 73, at 77: the tenant's fraction 'is calculated as being, broadly speaking, that proportion of the total of the factors which go into milk production which the tenant's dairy improvements and fixed equipment contribute.'

[133] The Agriculture Act 1986, Sch.1, para.5(2)(a)(i), (b), and (c).

[134] Ibid., para.7(1)(a).   [135] Ibid., para.7(1)(b).

improvements and fixed equipment is taken to be the amount which, on a
rent arbitration under the AHA 1986, would fall to be disregarded under
paragraph 2(1) of Schedule 2 to that Act 'so far as that amount is attributable to tenant's improvements to, or tenant's fixed equipment on, land used
for the feeding, accommodation or milking of dairy cows kept on the land
in question'.[136] In turn, this provison is modified to take account of the fact
that the tenant's fraction is calculated by reference to the relevant period —
unlike assessment of compensation for improvements under the AHA
1986, which addresses the position as at the termination of the tenancy. For
example, allowances made or benefits given by the landlord in consideration of the execution of improvements wholly or partly executed at the
tenant's expense fall to be disregarded if made or given after the end of the
relevant period.[137] A further modification is that, where a successor tenant
takes over his predecessor's entitlement to compensation under the Agriculture Act 1986,[138] the successor enjoys the benefit of that predecessor's
improvements and fixed equipment.[139]

Certain aspects of the calculation may be highlighted. First, the provisions introduce a yet further alternative to 'areas used for milk production'
and land used 'for the feeding of dairy cows' — namely 'land used . . . for
the feeding, accommodation or milking of dairy cows kept on the land'.
This last expression would seem to occupy the middle ground between
the other two. For example, while land used for dairy or dual-purpose
bulls is excluded, it does cover certain 'accommodation'. Secondly, the
annual rental value of the tenant's dairy improvements and fixed equipment
is assessed at the *end* of the relevant period. However, recourse is had to
the rent payable *throughout* the relevant period when assessing, in the
case of the denominator, the proportion attributable to the land used in
that period for the feeding, accommodation, or milking of dairy cows
kept on the land.[140] Thirdly, the Court of Appeal in *Creear* v. *Fearon*
has thrown into sharp relief a further distinction to be made in this

---

[136] Ibid., para.7(2). In this context 'tenant's improvements' and 'tenant's fixed equipment'
have the same meaning as in para.2 of Sch.2 to the AHA 1986. See also n.52, *supra*.
[137] The Agriculture Act 1986, Sch.1, para.7(4)(a). By contrast, under para.2 of Sch.2 to the
AHA 1986 'tenant's improvements' do not include any improvements where an equivalent
allowance or benefit has been made or given by the landlord in consideration of their execution.
[138] I.e., under the Agriculture Act 1986, Sch.1, para.2.
[139] Ibid., para.7(4)(c). For a third modification, see ibid., para.7(4)(b). It may also be noted
that, for the purposes of calculating the denominator, provision is made for circumstances
where the relevant period is greater or less than twelve months or the rent was only payable
by the tenant in respect of part of the relevant period. In such circumstances it is necessary to
ascertain the average rent payable in respect of one month in the relevant period or, as the case
may be, in respect of that part of the relevant period; and then the figure to be inserted into the
calculation is taken to be the corresponding annual amount: ibid., para.7(3).
[140] The effect could be significant where there had been a change in the rent during the
relevant period.

context.[141] For the purpose of ascertaining the numerator and, accordingly, the first element of the denominator, determination of the annual rental value of the tenant's dairy improvements and fixed equipment is to be carried out, broadly, by calculating the amount which would fall to be disregarded on a rent arbitration, so far as attributable to tenant's improvements to, and tenant's fixed equipment on, land used for the feeding, accommodation, or milking of dairy cows kept on the land. There is no requirement to identify the actual rental value of the unimproved land, nor should account be taken of the actual rent payable under the tenancy. However, whereas in these circumstances the focus is upon the *rental value*, in the case of the second element of the denominator the focus is on the proportion of the *rent payable* in respect of the relevant period attributable to the land used in that period for the feeding, accommodation, or milking of dairy cows kept on the land — i.e., the proportion of the actual rent. As happened on the facts of *Creear* v. *Fearon*, this can operate very much to a landlord's disadvantage if by concession he had permitted the tenant to pay a low rent.[142] The lower the rent, the smaller the second element of the denominator; and the smaller the second element of the denominator, the greater the tenant's fraction. Fourthly, and lastly, since the calculation is made by reference to the relevant period (which, as has been seen, for most wholesale producers will be the 1983 calendar year and for most direct sales producers the 1981 calendar year), the advantage of possessing full historic records cannot easily be overestimated. In the event of dispute, much may turn on the extent and accuracy of such evidence.

4.2.4.4. Transferred quota

'Transferred quota' is defined as 'milk quota transferred to the tenant by virtue of the transfer to him of the whole or part of a holding';[143] and the tenant receives compensation for such quota to the extent that he bore the costs. Accordingly, if he bore the whole of the cost of the transaction by virtue of which the quota was transferred, he is entitled to its full value; but, if he only bore a part of the cost, his entitlement is limited to the value of the corresponding part.[144] Again care must be taken to ensure that the tenant pays the cost of the transaction rather than, for example, a family company in respect of which he is a director and/or shareholder. At the same time it may be re-emphasized that to be eligible for such a claim the tenant need

---

[141] [1994] 46 EG 202.

[142] In *Creear* v. *Fearon* the tenant paid an annual rent of only £400 during the relevant period (the 1983 calendar year) for a farm of approximately 101 acres.

[143] The Agriculture Act 1986, Sch.1, para.1(2). This definition reflects the general requirement for a land transaction to transfer milk quota and excludes payment for any quota temporarily transferred into the name of the tenant as at the termination date.

[144] Ibid., para.5(3). As with transferred quota to be disregarded on a rent review, there is the need for a transaction.

not have had quota allocated to him; but, if he did not, as a rule he must have been in occupation as at 2 April 1984.[145]

4.2.5. Valuation

Once the amount of any 'excess over standard quota', 'tenant's fraction', and 'transferred quota' has been ascertained, it remains to determine the value of the quota, the Agriculture Act 1986 expressly recognizing this economic reality. Paragraph 9 of Schedule 1 states that the value in question is that at the termination of the tenancy. Further, it directs that in determining the value 'there shall be taken into account such evidence as is available, including evidence as to the sums being paid for interests in land — (a) in cases where milk quota is registered in relation to the land; and (b) in cases where no quota is so registered'. This short provision, largely addressing matters of evidence, has had widespread implications. Most notably, it has proved a forum for debate as to the validity of quota transfers operated through the medium of short-term agreements, tenants being eager to adduce details of this category of transaction where prices paid have historically been high.

In particular, the effect of these valuation provisions were considered in *Carson* v. *Cornwall County Council*.[146] Five methods of valuation were canvassed before the arbitrator. The first was based upon prices paid for quota transferred through the medium of short-term agreements;[147] the second on prices paid for temporary transfers of quota; the third on prices paid for vacant possession land sold respectively with and without quota;[148] the fourth on rents paid for land let respectively with and without quota; and the fifth on comparable compensation settlements. The arbitrator felt that only the second and third methods provided an acceptable basis for determining the value of compensation. In particular, with regard to the first method, he accepted Professor Usher's expert advice that there was doubt whether grazing tenancies or licences had the effect of transferring quota for the purposes of Article 7 of Council Regulation (EEC) 857/84,[149] whether tenancies or licences of less than a marketing year in duration

---

[145] There is also the possibility that he could benefit from the provisions in favour of certain successors, assignees, and head-tenants: ibid., paras.2–4.

[146] [1993] 03 EG 119. See also Densham, H. A. C., *Scammell and Densham's Law of Agricultural Holdings, 1993 Supplement*, at A20–1. The decision is of the greater interest in that both the County Court and earlier the arbitrator had the benefit of Professor Usher's expert advice.

[147] As at the date of termination (29 September 1987) either tenancies or licences were apprehended to be capable of triggering a transfer of quota. See Ch.4, 3.

[148] It is arguable that this alternative should receive greater weight in that it mirrors most closely the wording of the statute.

could have such effect, and whether land that producers were expressly prohibited from using for milk production could constitute part of a 'holding' within the Community regulations.[150] The Court dismissed the appeal. In so doing it emphasized that compensation under the Agriculture Act 1986 was payable in respect of the tenant's efforts in building-up milk production on the holding, with the result that it would be wrong to give weight to prices paid in a 'quota market', where producers (dependent on their own individual circumstances) might be paying 'above the odds'. By way of illustration, prices paid for both permanent and temporary transfers are likely to become unrepresentatively high towards the close of milk years in which the superlevy is triggered, with individual producers who have exceeded their reference quantities being prepared to pay large sums for the acquisition of quota to limit their liability. Accordingly, considerable injustice would occur if so much was allowed to depend upon whether at the critical date, the termination of the tenancy, the prospective payment of superlevy had led to the inflation of prices.[151]

4.2.6. Miscellaneous

In addition, there is statutory provision governing various ancillary matters, many of them procedural. First, in certain circumstances a tenant is entitled to claim compensation when his landlord resumes part only of the tenanted land.[152] An apportionment (presumably in accordance with the areas used for milk production) will be required to ascertain the amount of quota referable to the land vacated and the amount referable to the land remaining subject to the tenancy.[153]

Secondly, where the reversion is vested in more than one person in several parts, the tenant is entitled to require that compensation be assessed on the basis that the reversion were not so severed. This should secure for

---

[149] [1984] OJ L90/13.

[150] For more detailed consideration of these aspects, see Ch.1, 5; and Ch.4, 4.1. and 4.2.

[151] In an agreement between Gloucestershire County Council and certain of its former tenants, reported at [1993] 43 EG 55, an 'amended market value' approach was adopted, so as to strip out any enhancement in prices arising through fear of superlevy.

[152] The AHA 1986, Sch.1, para.13. The specified circumstances are: (1) where the landlord resumes possession of part of the tenanted land having served notice to quit part under the AHA 1986, ss.31 or 43(2); (2) where the landlord resumes possession of part of the tenanted land in accordance with a contractual provision; and (3) where a notice to quit part is given by a person entitled to a severed part of the reversion, by virtue of the Law of Property Act 1925, s.140 (on which, see, e.g., Muir Watt, J., *Agricultural Holdings*, at 99–104; Densham, H. A. C., *Scammell and Densham's Law of Agricultural Holdings*, at 199–203; and Rodgers, C. P., *Agricultural Law*, at 110–13).

[153] This matter would become more intricate if the tenant operated a Euro-holding comprising land over and above that subject to the tenancy.

the tenant any advantage attributable to the quota being presented on the market as a whole rather than in separate lots.[154] The arbitrator is directed, where necessary, to apportion the amount awarded between the owners of the reversion; and any additional costs caused by the apportionment are to be paid proportionately by those persons in accordance with the arbitrator's determination.[155]

Thirdly, the legislation addresses circumstances where the landlord does not own the freehold in fee simple (or, in the event that the landlord is himself a tenant, where he does not own that leasehold interest absolutely).[156] In such circumstances, for the purposes of claims for compensation under the Schedule, the landlord may do anything which he might do if he were a fee simple owner (or absolutely entitled to the leasehold interest).

Fourthly, the compensation provisions are expressly applied to land belonging to Her Majesty in right of the Crown or to the Duchy of Lancaster, the Duchy of Cornwall, or a Government department; or to land which is held in trust for Her Majesty for the purposes of a Government department.[157]

Fifthly, just as there may be prospective apportionment in the case of quota transfers, it is open to the parties to ascertain in advance certain, key factors in the calculation of the compensation claim. Either party prior to the termination of the tenancy, may require by notice in writing that the determination of the standard quota for the land or the tenant's fraction be referred to arbitration. The arbitrator on such a referral is directed to determine the standard quota for the land or, as the case may be, the tenant's fraction (so far as determinable at the date of the reference).[158] Without doubt this facility has considerable advantages. Not only is the requisite evidence likely to be more contemporaneous; but as from the date of the advance determination both parties will be aware of two key factors liable to influence the size of any subsequent claims. However, the value of the quota, which could equally prove to be a key factor, must remain unascertained until the termination of the tenancy. Should the standard

[154] The AHA 1986, Sch.1, para.14(1). Cf., the AHA 1986, s.75. For discussion in the tax context of sales by lot as opposed to sales as a whole, see, e.g., *A-G of Ceylon* v. *Mackie*, [1952] 2 All ER 775; *Duke of Buccleuch* v. *I.R.C.*, [1967] 1 AC 506; and *Gray* v. *I.R.C.*, [1994] STC 360, [1994] 38 EG 156.

[155] The AHA 1986, Sch.1, para.14(2).

[156] Ibid., para.15. Cf., the AHA 1986, s.88. The terminology is not easy to equate with the position following the 1925 law of property legislation. In particular, a freehold estate may now only exist as a fee simple absolute in possession.

[157] Ibid., para.17.

[158] Ibid., para.10(1) and (2). The legislation seems to contemplate the determination of *either* the standard quota *or* the tenant's fraction. On a strict reading it does not seem to countenance determination of both (at least under the same arbitration). This would, in all likelihood, be contrary to the wishes of the parties. On this aspect, see Densham, H. A. C., *Scammell and Densham's Law of Agricultural Holdings*, at 334.

quota or tenant's fraction be so determined, it is provided that, where the matter is subsequently referred to arbitration, the arbitrator must make his award in accordance with the determination —— unless it appears that any relevant circumstances were materially different at the point when the tenancy ended from those at the date of the determination. If that is the case, the arbitrator is to disregard so much of the determination as appears to him to be affected by the change in circumstances.[159] It is also of note that, unlike prospective apportionments under the transfer provisions, which are only effective for six months from any arbitration, any advance determination of the amount of standard quota or of the tenant's fraction is effective until the termination of tenancy, whenever that may be.[160]

Sixthly, a very similar provision permits a landlord and a tenant, prior to the termination of the tenancy, to agree in writing the amount of the standard quota for the land or the tenant's fraction or, in this case, the value of the quota without resort to arbitration. No guidance is given as to whether this agreement must be made in accordance with the terms of the statute or whether the parties are at liberty to adopt their own criteria. Again the agreement is binding where on termination of the tenancy the matter is referred to arbitration — unless it appears to the arbitrator that there has been a material change of circumstances.[161]

Finally, Schedule 1 lays down detailed procedures for the conduct of any arbitration. Those instigated for the purposes of advance determination of the standard quota or tenant's fraction are to be conducted in accordance with section 84 of the AHA 1986 (which brings into play Schedule 11 of the AHA 1986, *mutatis mutandis*).[162] Arbitrations on the claim for compensation itself are again to be conducted in accordance with the provisions of the AHA 1986, albeit with modification.[163] In these circumstances it is critical that the tenant serve on his landlord within two months of the termination of the tenancy notice in writing of his intention to make a claim. Failure to do so renders the claim unenforceable.[164] The two month time-limit is the same as that applicable under section 83(2) of the AHA 1986; but, unlike section 83(3) of the AHA 1986, there is no requirement that the tenant 'specify the nature of the claim'. An illustration of the latitude enjoyed by the tenant may be seen in the County Court case of *Walker* v.

---

[159] The Agriculture Act 1986, Sch.1, para.11(6) and (7).

[160] Subject only to a material change in circumstances: ibid., para.11(7).

[161] Ibid., para.11(6) and (7). See also Rodgers, C. P., *Agricultural Law*, at 335.

[162] The Agriculture Act 1986, Sch.1, para.10(3). On arbitrations under the AHA 1986, generally, see, e.g., Muir Watt, J., *Agricultural Law*, at 273–317; Densham, H. A. C., *Scammell and Densham's Law of Agricultural Holdings*, at 363–86; and Rodgers, C. P., *Agricultural Law*, at 359–75.

[163] The Agriculture Act 1986, Sch.1, para.11. Likewise, arbitrations relating to the apportionment of quota largely follow the AHA 1986 procedure: the Dairy Produce Quotas Regs. 1994, S.I.1994 No.672, Sch.2 (in the case of England and Wales).

[164] The Agriculture Act 1986, Sch.1, para.11(1).

Crocker.[165] A letter had been written to the landlords stating, *inter alia*, that: 'if your clients are successful on November 20 [the date of the possession hearing, in the event adjourned generally] then our client expects compensation for all his input into the farm including the value of the (milk) quota'. This was held to constitute a valid notice for the purposes of the claim for milk quota compensation — and, it may be noted, also for the purposes of section 83 of the AHA 1986.

Following termination, the parties receive an eight month period in which to reach agreement in writing as to the amount of compensation; and, in default of agreement, the claim must be determined by arbitration under the AHA 1986.[166] However, where the tenant lawfully remains in occupation of part of the land subject to the tenancy following termination, neither the two month nor the eight month periods begin to run until the termination of that occupation.[167]

As indicated, if the compensation claim actually proceeds to arbitration, the AHA 1986 procedure is employed, subject to modification. Whereas under the AHA 1986 the arbitrator's award must fix a day for payment not less than one month after the delivery of the award, in the case of a milk quota compensation claim this period may be extended for as long as three months from the award.[168] Further, as also seen, in the context of milk quota compensation the arbitrator may be constrained by an advance determination or agreement between the parties as to the standard quota for the land or the tenant's fraction.

For the purposes of securing payment, recourse may again be had to the AHA 1986. Both the enforcement provisions and certain of those relating to a landlord's ability to obtain a charge on the holding are expressly applied to sums due under Schedule 1 of the Agriculture Act 1986.[169] Finally, *mutatis mutandis*, the rules governing the service of notices accord with those under section 93 of the AHA 1986. In particular, any notice is duly served on the person to be served if delivered to him, or left at his proper address, or sent to him by post in a registered letter or by recorded delivery.[170]

---

[165] [1992] 23 EG 123.

[166] The Agriculture Act 1986, Sch.1, para.11(2). Again this period is the same as under the AHA 1986: the AHA 1986, s.83(4).

[167] The Agriculture Act 1986, Sch.1, para.11(4). Cf., the AHA 1986, s.83(6). The general rules governing time limits are also adjusted to take account of circumstances where a new tenancy of the land or part of the land may be granted to a different tenant by virtue of a direction under s.39 of the AHA 1986: the Agriculture Act 1986, Sch.1, para.11(3) and (4).

[168] The Agriculture Act 1986, Sch.1, para.11(5). Cf., the AHA 1986, Sch.11, para.18. Presumably the purpose of this extension is to take account of the large sums which the landlord may be required to find.

[169] The Agriculture Act 1986, Sch.1, para.12. For the AHA 1986 provisions applicable, see s.85 (enforcement) and s.86(1), (3), and (4) (power of landlord to obtain charge on holding).

[170] The Agriculture Act 1986, Sch.1, para.16.

4.3. Compensation outside the Agriculture Act 1986

While the Agriculture Act 1986 does grant certain tenants an extensive entitlement to compensation, the limitations are at once apparent. For example, with the passage of time there are now many tenants who cannot demonstrate that they received a quota allocation. In particular, allocations under the general rule were made to those who were actively involved in milk production over a decade ago, whether the quota be wholesale quota or direct sales quota.[171] A degree of latitude is granted in respect of transferred quota, in that such a claim may be sustained notwithstanding that the tenant did not receive a quota allocation. That having been said, in the absence of a quota allocation the tenant must as a rule have been in occupation as tenant on 2 April 1984.[172] Consequently, the combination of these and the other provisions governing eligibility present a formidable hurdle; and, moreover, an influential factor could be farming practices adopted at a time when the availability of quota compensation was far from the tenant's mind. By way of illustration, the tenant may have been operating a system of extensive grazing during the relevant period, so prejudicing the claim for excess over standard quota. Finally, it has also been seen that tenants who vacated prior to 25 September 1985 received no statutory entitlement to compensation; and that the ATA 1995 expressly excludes compensation under the Agriculture Act 1986 in the case of farm business tenancies.

Accordingly, many tenants will not qualify under the Agriculture Act 1986, and there can be little doubt that they are placed at a considerable disadvantage. In the case of those who are still governed by the AHA 1986, their only recourse is to secure agreement with their landlords. In particular, appropriate terms should be obtained if the tenant provides the consideration for a permanent transfer of quota onto the tenanted land. For example, the tenant may seek the right to enter into a permanent transfer of such quota from the tenanted land during the currency of the agreement. In this connection he should also seek agreement that consent will be forthcoming to the grant of a subtenancy of at least ten months duration in order to comply with the statutory requirements for a change of occupation.[173]

---

[171] For most wholesale producers the UK reference year was the 1983 calendar year; and for most direct sales producers it was the 1981 calendar year. Although secondary allocations may also confer entitlement, these again are grounded in historic circumstances. However, as indicated, in the case of certain successors, assignees, and head-tenants compensation in respect excess over standard quota and the tenant's fraction may yet be available even where the claimant himself received no allocation: the Agriculture Act 1986, Sch.1, paras.2–4.

[172] As with allocated quota, the general rule is tempered in the case of certain successors, assignees, and head-tenants: ibid.

[173] In this context it would seem to be common practice for the landlord and tenant to agree that, before any transfers of quota from the tenanted land, the landlord should have the right to purchase the quota himself at market price.

In the case of farm business tenants under the ATA 1995, their rights are more extensive. Although the Agriculture Act 1986 is not applicable, the ATA 1995 (unlike the AHA 1986) provides that tenant's improvements include any 'intangible advantage' which is 'obtained for the holding by the tenant by his own effort or wholly or partly at his own expense'; and which 'becomes attached to the holding'.[174] It is generally understood that this category of improvement embraces milk quotas.[175] However, great care should be taken by tenants who move from the AHA 1986 régime to the farm business tenancy régime. Where the AHA 1986 tenancy gives rise to a statutory right to compensation under the Agriculture Act 1986, that statutory right cannot be carried over to the farm business tenancy. Accordingly, it is essential to crystallize the claim on the termination of the earlier AHA 1986 tenancy.[176]

Further, eligibility for compensation is dependent on landlord's consent in writing.[177] Such consent may be found in the tenancy agreement or elsewhere, and it may be given at any time during the currency of the farm business tenancy. That having been said, the landlord may impose conditions or grant consent subject to the tenant's acceptance of a specified variation in the terms of the tenancy — although such must relate to the improvement in question.

While the emphasis is on agreement between the parties, the ATA 1995 does in certain circumstances offer the alternative of demanding by notice in writing that the matter be referred to arbitration.[178] If the arbitrator's approval is obtained, it is treated for the purposes of the compensation provisions and the terms of the farm business tenancy as if it were landlord's consent. In making his award, the arbitrator is directed to consider whether it is reasonable for the tenant to provide the improvement, having regard to the terms of the tenancy and other relevant circumstances, including the circumstances of the tenant and the landlord (which would seem to include the landlord's ability to pay the heavy sums which may become due). This ability to go to arbitration undoubtedly mitigates the tenant's position. A

[174] S.15(b). For these purposes 'holding' means the land comprised in the farm business tenancy rather than the tenant's Euro-holding. On compensation for improvements under the ATA 1995, generally, see, e.g., Evans, D., *The Agricultural Tenancies Act 1995*, at 8/15–27; Moody, J., *Agricultural Tenancies Act 1995 — a Practical Guide*, at 29–36; and Sydenham, A. and Mainwaring, N., *Farm Business Tenancies: Agricultural Tenancies Act 1995*, at, in particular, 66–81.

[175] For fuller discussion of this aspect, see nn.59–61, *supra*.

[176] See, e.g., Evans, D., *The Agricultural Tenancies Act 1995*, at 8/16; and Sydenham, A. and Mainwaring, N., *Farm Business Tenancies: Agricultural Tenancies Act 1995*, at 70.

[177] The ATA 1995, s.17. This section governs the eligibility criteria for all physical improvements and for all intangible advantages except planning permissions.

[178] S.19. The tenant may seek arbitration if aggrieved by the refusal of his landlord to give consent; by the failure of his landlord to give consent within 2 months of a written request; or by any variation in the terms of the tenancy required by the landlord as a condition for consent.

further provision which may operate to the tenant's benefit is that the arbitrator must either grant approval or withhold it, there being no power to impose conditions or to vary the terms of the tenancy.[179] Accordingly, by way of illustration, if the transfer of quota onto the tenanted land were effected pursuant to arbitrator's approval, the landlord would seem unable to insist on the inclusion of quota clauses. By contrast, the tenant is faced by a very significant constraint in that as a general rule he must demand arbitration *before* he has begun to provide the improvement.[180] Routine improvements are an exception to the general rule; but they would not seem apt to include the transfer of milk quota. In addition, any covenant against the making of improvements without consent is not subject to a statutorily imposed requirement that consent must not be unreasonably withheld, section 19 of the Landlord and Tenant Act 1927 being inapplicable.[181]

Should compensation become payable, the amount due (other than in the case of planning permissions) is that 'equal to the increase attributable to the improvement in the value of the holding at the termination of the tenancy as land comprised in a tenancy'.[182] Much will again depend on the value of quota prevailing as at the termination date.[183]

It may also be noted that, as with rent review, the AHA 1986 arbitration procedure is not applicable. Instead, any claim for compensation in respect of tenant's improvements is to be determined under the general law governing arbitrations.[184] This general law also governs disputes as to consent for improvements. There are, however, certain additional provisions which the ATA 1995 expressly applies. In particular, detailed rules govern the time limits in respect of both disputes as to consent for improvements and in respect of the claim for compensation on termination. A tenant must demand arbitration as to consent for improvements by notice in writing to the landlord within two months from receipt of the landlord's refusal; or within

[179] Ibid., s.19(6). If the arbitrator withholds approval where the tenant has disputed the variation required by the landlord as a condition for consent, then the landlord's consent and the condition are valid: ibid., s.19(8).

[180] Ibid., s.19(2). In theory the tenant may obtain by agreement his landlord's written consent in writing at any time during the currency of the farm business tenancy, without recourse to arbitration; but in practice his negotiating position will be substantially weakened as soon as he begins the improvement without consent.

[181] See now the ATA 1995, Sch., para.6.

[182] Ibid., s.20(1). S.20(2) provides that: 'Where the landlord and the tenant have entered into an agreement in writing whereby any benefit is given or allowed to the tenant in consideration of the provision of a tenant's improvement, the amount of compensation otherwise payable in respect of that improvement shall be reduced by the proportion which the value of the benefit bears to the amount of the total cost of providing the improvement.'

[183] There is a marked contrast with the complex machinery of the Agriculture Act 1986; but in part this may be regarded as a product of the relative importance of allocated quota under the earlier legislation.

[184] The ATA 1995, s.22(1) — i.e., recourse is had to the Arbitration Acts 1950–79.

two months from receipt of the landlord's notice requiring a variation of the terms of the tenancy; or, if the landlord fails to give consent within two months from a written request, within four months from the written request.[185]

In the case of the claim for compensation itself, the tenant must within two months from the date of termination serve a written notice on his landlord of his intention to make the claim and of the nature of the claim.[186] This requirement is in line with the two month limit under both the AHA 1986 and the Agriculture Act 1986; but, unlike the Agriculture Act 1986, the tenant must specify in terms the nature of the claim. Further, as with both the AHA 1986 and the Agriculture Act 1986, an extension is granted where the tenant lawfully remains in occupation of part of the tenanted land after the termination of the farm business tenancy.[187] Finally, there are supplementary rules which address, *inter alia*, successive tenancies,[188] circumstances where the landlord resumes possession of part of the tenanted land,[189] circumstances where the reversion has been severed,[190] and the extent to which compensation may be recovered by agreement.[191] The last of these is perhaps somewhat surprising, in that where the statutory compensation scheme applies it overrides any agreement to the contrary, notwithstanding the emphasis on freedom of contract which pervades the ATA 1995.

Accordingly, it would appear that the farm business tenant must to a considerable extent rely on agreement with his landlord if he is to secure compensation in respect of milk quota.[192] While there is the possibility of demanding arbitration, such arbitration is subject to considerable limitations, not least that it is too late to seek approval once the improvement has been commenced. There would seem to be no substitute, therefore, for ensuring that such issues are addressed comprehensively at the outset in the

---

[185] The ATA 1995, s.19(3). See also s.19(9), which addresses circumstances where a tenant begins to provide an improvement (other than a routine improvement) after giving a notice demanding arbitration as to consent. For provisions as to the appointment of the arbitrator, see s.19(4).

[186] Ibid., s.22(2). For provisions as to the appointment of the arbitrator, see s.22(3).

[187] Ibid., s.22(5).

[188] Ibid., s.23. Cf., the Agriculture Act 1986, Sch.1, para.2. However, the successive tenancy provisions will not assist a tenant who gives up a tenancy protected by the AHA 1986 which carries with it entitlement to compensation under the Agriculture Act 1986 and takes on a new tenancy under the ATA 1995. For this reason such a tenant must insist that he is paid the compensation on termination of the AHA 1986 tenancy.

[189] The ATA 1995, s.24. Cf., the Agriculture Act 1986, Sch.1, para.13.

[190] The ATA 1995, s.25. Cf., the Agriculture Act 1986, Sch.1, para.14.

[191] The ATA 1995, s.26.

[192] As indicated, agreement between the parties is expressly countenanced by Council Reg. (EEC) 3950/92, [1992] OJ L405/1, Art.7(2). See Ch.5, 1. None the less, in the absence of agreement, the same Article requires that account be taken of the legitimate interests of the parties.

tenancy agreement itself. In particular, account could be taken of any quota transferred into the name of the tenant on moving into occupation; and of any quota which he may subsequently bring onto the tenanted land (whether by purchase from third parties or, in the event that he farms a composite holding, by change in the areas used in milk production). At the same time the landlord would enjoy the opportunity to protect his position by the full array of quota clauses already considered.[193] Failure to include appropriate terms in the tenancy agreement must almost inevitably work to the disadvantage of one party or the other. A prime example is provided by the situation where a tenant of a composite holding over a period of time switches his milk production from other land which he occupies, whether as freehold owner or as tenant, to the land comprised in the farm business tenancy. Unless he has obtained his landlord's written consent or demanded arbitration prior to the commencement of the improvement, he will not be able to sustain a claim for compensation.

---

[193] In this regard it may be reiterated that the parties cannot contract out of the compensation provisions in the ATA 1995: the ATA 1995, s.26.

6

Taxation

1. INTRODUCTION

Just as the nature of milk quota is itself elusive, so the taxation of milk quota gives rise to few easy solutions.[1] Not least, tax considerations were unlikely to have been at the forefront of the legislature's mind when implementing the milk quota system. In any event, it was not the intention of the legislature that quota should be so freely bought and sold or, indeed, acquire so transparent an economic value. Likewise, the judgments of the courts have not as a rule been directed to fiscal matters; and there may be dangers in seeking to derive guidance as to the tax treatment of quota from interpretations of the legislation delivered in other contexts — and vice versa. In this regard may be cited the decision of the High Court in *Faulks* v. *Faulks* and the decision of the Special Commissioners in *Cottle* v. *Coldicott*.[2] The former, as has already been seen, concerned the treatment of milk quota on the termination of a partnership. In holding that the milk quota passed with the tenancy to the surviving partner, Chadwick J rejected arguments to the contrary grounded in the tax legislation. More specifically, he did not feel it right that the status of milk quota as a separate asset for the purposes of Capital Gains Tax roll-over relief should disturb his detailed analysis of the primary regulations. By contrast, in *Cottle* v. *Coldicott* the Special Commissioners held that, for the purposes of the Capital Gains Tax legislation, milk quota would constitute a separate asset when transferred through the medium of short-term tenancies. However, emphasis was laid, *inter alia*, upon the fiscal context and upon the fact that the main objective of the transaction was the transfer of quota. As a result the Special Commissioners could expressly affirm that their decision was not inconsistent with

---

[1] For discussion of the taxation of milk quotas, generally, see, e.g., Hill, S. O., 'Milk Quotas: Legal and Taxation Implications', L.S.G. 20 Jan. 1988, 19–20; Fitzpatrick, F., 'Milk Quota: a Hybrid Animal', (1993) 207 Tax Journal 12–13; Williams, D. and Paskins, H., 'Milk Quota', (1993) 207 Tax Journal 13–15 and 208 Tax Journal 13–15; Harris, D., 'Sorting the Sheep from the Cows', (1993) 221 Tax Journal 14–16; Cardwell, M. N. and Lane, S., 'The Taxation of Milk Quota', [1994] B.T.R. 501–26; Snape, J., 'Transfers of Milk Quotas: Law and Tax', (1995) 2 Private Client Business 150–161; and Stanley, O., *Taxation of Landowners and Farmers* (Looseleaf edn., Butterworths, London, 1995), at 2.66.

[2] Respectively, [1992] 15 EG 82; and [1995] SpC 40. See also Ch.4, 2.1. Moreover, in *Harries* v. *Barclays Bank plc* the Court expressly viewed these decisions as difficult to reconcile: (not yet reported, High Court, 20 Dec. 1995).

approach adopted in *Faulks* v. *Faulks*, where the application of the general rule was appropriate.

Many of the difficulties flow from the relationship between milk quota and land. As indicated, it has throughout remained the general principle that milk quota is 'linked' with land; but due recognition must also be given to the fact that the general principle is now subject to significant exceptions. Where the United Kingdom tax legislation draws a distinction between land and other forms of asset, the strength of the link will be critical to the form of assessment. Further, it may also be necessary to accept that, as the strength of the link varies to meet diverging Community objectives, the tax treatment should likewise vary.

Accordingly, on the one hand, tax issues have highlighted many areas of difficulty; but, on the other hand, the necessary rigour required to address those issues has shed considerable light on the operation of the milk quota system as a whole. With producers entering into an large numbers of transactions year after year, it has not been possible to avoid consideration of the correct form of charge; and it may, therefore, be no coincidence that much of the debate as to the nature of milk quota has been conducted in the fiscal context.

Four charges to tax may be considered: Capital Gains Tax, Inheritance Tax, Income Tax, and Value Added Tax.

2. CAPITAL GAINS TAX

2.1. General

The Capital Gains Tax treatment of milk quota has attracted perhaps greatest attention, proving the cockpit of dispute between taxpayers and the Inland Revenue as to the fiscal consequences of the link with the land or holding. In particular, taxpayers have resisted the contention that quota constitutes a separate asset, an interpretation which would normally operate very much to their disadvantage. For example, if milk quota is indeed a separate asset, any allocations received when the system was implemented on 2 April 1984 would apparently lack a base cost, since the acquisition was not by way of a bargain at arm's length and there was probably no corresponding disposal.[3] Further, where there is no base cost, indexation will be also unavailable. Indexation may yet be available on the land or holding; but the scope for using this to shelter a gain on the quota is now subject to the provisions in the Finance Act 1994 which (subject to transitional relief) preclude the use of indexation to create or increase a capital loss.[4] By way of final example, no roll-over relief will have been

---

[3] The Taxation of Chargeable Gains Act 1992, s.17. The effect in tax law of the imposition of milk quotas was considered by the Special Commissioners in *Cottle* v. *Coldicott* (with regard to what is now the Taxation of Chargeable Gains Act 1992, s.43). See further Ch.6, 2.3.

[4] The Finance Act 1994, s.93 and Sch.12.

available until authorized by specific statutory provision effective from midnight, 29 October 1987.[5]

On the other hand, if milk quota is correctly to be regarded as an interest in or right over the land or holding, then its permanent transfer would constitute a part disposal of that land or holding. Further, *ab initio* it should have been a qualifying asset for the purposes of roll-over relief; and indexation would be applied to the single asset, a notable advantage where the taxpayer enjoyed the benefit of 31 March 1982 rebasing.[6]

However, before considering whether milk quota is to be viewed as a distinct and separate asset for Capital Gains Tax purposes, two preliminary matters may be considered. First, there has been debate as to whether it is an asset at all. In *Wachauf* v. *Bundesamt für Ernährung Forstwirtschaft* the United Kingdom Government and the Commission took the stance in their written observations that milk quota was merely an instrument of market management, with the result that it could not be a kind of intangible asset in which property rights might arise.[7] It may also be argued that, in the course of his judgment in *Faulks* v. *Faulks*, certain dicta of Chadwick J on the nature of temporary transfers of quota suggest that these may be better regarded as transfers of potential liability to superlevy rather than as transfers of property (the transferors' potential liability being increased and the transferees' potential liability being correspondingly diminished). However, as shall be seen, temporary transfers throw up their own tax problems; and in any event his primary concern was to demonstrate that milk quota is not a separate asset. Subsequently, in *Von Deetzen II* and *R.* v. *Ministry of Agriculture, Fisheries and Food*, ex parte *Bostock* the European Court took the view that the right to property enshrined in the general principles of Community law did not include 'the right to dispose, for profit, of an advantage', such as milk quota allocated in the framework of the common organisation of a market, 'which does not derive from the assets or occupational activity of the person concerned'.[8] That having been said, it may be observed that in both cases fiscal considerations were again not uppermost in the mind of the Court; and, besides, both decisions emphasized that the Community legal order would not protect *the right to dispose* of milk quota for profit, milk quota itself being more neutrally described as an 'advantage'.[9] Further, in light of the high prices paid for quota, there would be

[5] For the current legislation, see the Taxation of Chargeable Gains Act 1992, s.155.

[6] For illustrations of the difference in the charge to tax depending upon whether quota is treated as an interest in or right over the land or holding, as a separate asset, see, e.g., Fitzpatrick, F., 'The Legal Nature of Milk Quota', (1992) 185 Tax Journal 4–5; and Williams, D. and Paskins, H., 'Milk Quota', (1993) 207 Tax Journal 13–15.

[7] Case 5/88, [1989] ECR 2609, [1991] 1 CMLR 328.

[8] Respectively, Case 44/89, [1991] ECR I–5119, [1994] 2 CMLR 487, at I–5156 (ECR) and 511 (CMLR); and Case 2/92, [1994] ECR I–955, [1994] 3 CMLR 547, at I–984 (ECR) and 570 (CMLR).

[9] On this aspect see the submissions of counsel for the taxpayer in *Cottle* v. *Coldicott*, [1995] SpC 40.

considerable unrealism in maintaining that it fell outside the tax net altogether. Indeed, it may be argued that transferability is a hallmark of property.

Secondly, if it is to be argued that quota is not a separate asset but rather an interest in or right over another asset, then the question arises as to whether that other asset is 'land' in the traditional sense or the 'holding' created by Community law. This question was highlighted in the recent decision of *Cottle* v. *Coldicott*.[10] According to the Special Commissioners the Community regulations indicated that the quota would correspond to the holding as opposed to individual parcels of land. Yet there was keen appreciation that such an interpretation could lead to conceptual difficulties, it being likely that the Capital Gains Tax code would recognize those individual parcels of land as assets, not the holding.[11] However, they expressly stated that, since the conceptual difficulties were not raised at the hearing, they did not form a part of the decision.[12]

In this light, and on the basis that milk quota does fall within the Capital Gains Tax legislation, the following potential heads of charge would seem most deserving of consideration. First, there could be an actual disposal of milk quota as a separate and distinct asset, for these purposes 'asset' being widely defined in section 21 of the Taxation of Chargeable Gains Act 1992 so as to include all forms of property. Secondly, the milk quota could constitute an interest in or right over the land or the holding. In these circumstances, by way of illustration, the sale of milk quota through the medium of short-term tenancies would trigger a part disposal. Thirdly, rather than an actual disposal of the milk quota or of an interest in or right over the land or holding, there could be a deemed disposal under section 22 of the Taxation of Chargeable Gains Act 1992, on the basis that a capital sum has been derived from assets. Under this third alternative it is to be noted that there must be some *asset* from which the capital sum has been derived.[13]

2.2. Actual disposals

As indicated, the Revenue has consistently maintained that quota is a separate asset for Capital Gains Tax purposes, its sale amounting to an actual disposal.[14] This interpretation has now been adopted by the Special

[10] Ibid. For a recent analysis of the 'holding' in the context of the Landlord and Tenant Act 1954, see *Graysim Holdings Ltd.* v. *P. & O. Property Holdings Ltd.*, [1995] 3 WLR 854.

[11] The complex consequences of treating the 'holding' as the asset were illustrated by the Special Commissioners: e.g., if a farmer began to sell milk so that his farm became a 'holding', would that give rise to an acquisition for CGT purposes?

[12] That having been said, there is no doubt that the difficulties were perceived as very real.

[13] For a full survey of the alternatives, see again the decision of the Special Commissioners in *Cottle* v. *Coldicott*, [1995] SpC 40.

[14] See, e.g., *Tax Bulletin* of February 1993; and *Tax Bulletin* of May 1994.

Commissioners in *Cottle* v. *Coldicott* — although at the same time they imposed close constraints on the limits of their decision.[15] Most notably, they emphasized that their conclusion was reached within the specific context of the Capital Gains Tax legislation; and that much depended upon the individual facts of the case.[16] In the latter regard the appellant had transferred the quota through the medium of a short-term tenancy, himself being appointed contractor to discharge the obligations of the transferee under the tenancy. While no comment was passed as to the validity of the transaction, its form was stated to be a relevant consideration; and the presence of three separate agreements (the quota transfer agreement, the tenancy agreement, and the contractor's agreement) operated against the appellant. Besides, it was observed that the main objective of such transactions is to transfer the quota away from the holding.

The conclusion of the Special Commissioners was reached on the basis of a careful review of the Community regulations and the relevant judgments of the European Court and English courts. From these quota was deduced to be primarily the entitlement of a producer. Not least, it was allocated to a producer rather than to a holding, and allocation was made by reference to milk production rather than to the extent of the holding.[17] There was, however, acceptance that under the general principle quota is transferred with a holding or part of a holding — unless one of the authorized derogations applies.[18] On the facts of the appeal it was highlighted that the transaction concerned only a part of the holding together with a separate transfer of part of the quota;[19] and, as mentioned, that the purpose of the transaction was to ensure that the quota was transferred away from the land comprised in the short-term tenancy.

Accordingly, the Special Commissioners were of the view that, while quota should as a rule be transferred with the whole or that part of the holding to which it corresponds, this does not mean that it forms a part of the holding. Rather, on the facts and in the context of the Capital Gains Tax legislation, milk quota is a personal asset of the producer and 'incorporeal property' within what is now section 21(1)(a) of the Taxation of Chargeable Gains Tax Act 1992.[20]

In addition, the Special Commissioners expressly rejected argument that

[15] [1995] SpC 40. See also Greenfield, R., Capital Tax Planning, October 1995, 135–9.

[16] For these reasons the Special Commissioners could declare that in their opinion *Faulks* v. *Faulks* was correctly decided on its own facts.

[17] A similar approach was adopted by the German Administrative Court in *Re the Küchenhof Farm*, [1990] 2 CMLR 289.

[18] For discussion of the derogations from the general principle that quota is transferred with a holding or part of a holding, see Ch.4, 5.

[19] In this the appeal contrasted with *Faulks* v. *Faulks*, which concerned the transfer of an entire holding.

[20] The legislation in force as at the operative date was the Capital Gains Tax Act 1979, s.19(1)(a).

the disposal of the quota amounted to the disposal of an interest in or right over the appellant's holding. As a result they held the part disposal rules inapplicable. More specifically, they did not consider quota to be a restriction attaching to the holding or a right of immunity enjoyed by the holding.[21] In their view it would be more accurate to characterize quota as an advantage, with the common organization of the market in milk and milk products since 1968 conferring on producers a benefit which was restricted but not removed by the introduction of the superlevy in 1984.[22] Recognition of this state of affairs was found in the judgment of the European Court in *Von Deetzen II*,[23] the buoyant values of dairy farms, the provisions governing temporary transfers, and the provisions permitting the transfer of quota without land for the purposes of restructuring.[24] Further, it was noted that, so long as a producer remains within his individual reference quantity, he secures the benefit of the common organization of the market in milk and milk products, and to this must be attributed at least part of the value of quota. Thus, it would be incorrect to regard the prices paid for quota as paid simply for a right of immunity.

A matter of some importance is that the Special Commissioners were not persuaded that quota exhibits the same characteristics as an option (the grant of which would undoubtedly trigger the part disposal rules). Indeed, whether milk quota is considered a separate asset or an interest in or right over another asset, it could legitimately be argued that no convincing analogy has yet been advanced. Suggestions have included a licence (to produce milk) and planning permission. Support for the former may be derived from the Opinion of the Advocate-General in *Wachauf* v. *Bundesamt für Ernährung und Forstwirtschaft*, where milk quota was viewed as 'a form of licence to produce a given quantity of a commodity (milk) at a more or less guaranteed price without incurring a penalty'.[25] However, while there is clear force in this analogy, Chadwick J in *Faulks* v. *Faulks* has pointed out that there is no requirement for a licence to produce milk in England and Wales.[26] As a result it is in theory open to a producer to exceed his indi-

---

[21] For judicial expression of the view that a quota is treated as an amount to represent the exemption from levy, see *R.* v. *Ministry of Agriculture, Fisheries and Food*, ex p. *Cox*, [1993] 22 EG 111, [1993] 2 CMLR 917.

[22] Cf., *Grounds* v. *A-G of the Duchy of Lancaster*, [1989] 21 EG 73, at 74 *per* Glidewell LJ: 'although in realistic terms the quota is a limitation, in a sense it has become a right with a value'.

[23] Case 44/89, [1991] ECR I–5119, [1994] 2 CMLR 487. However, in that case the Council in its written observations took the view that the restrictions imposed on the applicant were a straightforward limitation on the use of real property.

[24] Council Reg. (EEC) 3950/92, [1992] OJ L405/1, Arst.6 and 8; and the Dairy Produce Quotas Regs. 1994, S.I.1994 No.672, Regs.13 and 15, as amended by the Dairy Produce Quotas (Amendment) (No.2) Regs. 1994, S.I.1994 No.2919.

[25] Case 5/88, [1989] ECR 2609, [1991] 1 CMLR 328, at 2630 (ECR) and 342 (CMLR).

[26] [1992] 15 EG 82. For his purposes he would seem to have been disregarding any licence required under the public health regulations.

vidual reference quantity by a wide margin, but in practice excess production of this order is avoided, the punitive rate of superlevy rendering it uneconomic. Accordingly, in the sense that a licence authorizes activity which would otherwise be unlawful, the categorization of quota as a form of licence would not seem completely apposite.[27]

Planning permission and quota also share certain characteristics. Thus, the ATA 1995 expressly provides that planning permission may qualify for compensation on the basis that it is an 'intangible advantage' which 'becomes attached' to the holding; and quota is generally apprehended to satisfy the same criteria.[28] That having been said, quota may be transferred both from one part of a holding to another in accordance with the areas used for milk production and between holdings, a quality not shared by planning permission.[29] Further, if it is contended that milk quota is a separate asset, such a comparison would be of little assistance, planning permission being treated for tax purposes as an integral part of the land. In addition, although both quota and planning permission are regarded as intangible advantages under the ATA 1995, which would suggest that they constitute species of identifiable property, in essence the legislation provides that on termination they pass with the land to the landlord subject to a right to compensation in specific circumstances.[30]

Accordingly, the drawing of such analogies does shed considerable light on the nature of quota. Yet there remain difficulties in escaping the conclusion that, whether or not it is a separate asset, any attempt to confine it within the existing categories of property is liable to fail.[31] Rather, it would seem to stand alone as a creation of statute to meet the varying contingencies addressed by the Community legislature; and for this reason there is a certain inevitability that it does not fit easily into a tax system constructed with no such entity in mind.

Although *Cottle* v. *Coldicott* does provide clear authority to the effect that milk quota is a separate asset for Capital Gains Tax purposes when transferred through the medium of short-term tenancies, there are dangers in underestimating the lingering force of the High Court decision of *Faulks* v. *Faulks*.[32] On the one hand, there can be no denying that it related to the

---

[27] Further, the potential nature of a producer's liability creates additional strain on the analogy. He may exceed his individual reference quantity but escape payment of superlevy where the UK as a whole does not exceed its national guaranteed total quantity and/or (in the case of wholesale quota) where the purchaser which the producer supplies does not exceed its reference quantity. Thus, even if the producer goes beyond the terms of the 'licence', he may yet enjoy immunity.

[28] See the ATA 1995, s.15(b); and, generally, Ch.5, 4.3. However, as seen, *Cottle* v. *Coldicott* raises questions as to the extent to which quota satisfies this definition.

[29] This point was specifically mentioned in *Cottle* v. *Coldicott*.

[30] Comparisons may also be drawn with tax exemptions and reliefs; but again tax exemptions and reliefs as a rule lack the inherent characteristic of transferability.

[31] See, e.g., Rodgers, C. P., *Agricultural Law*, at 320–1.   [32] [1992] 15 EG 82.

treatment of milk quota on the dissolution of a partnership rather than a specifically fiscal context. On the other hand, it too contains a carefully reasoned analysis of the relationship between quota and the holding, with Chadwick J reaching the conclusion that quota 'could not properly be regarded as property, or a right or interest in property, separate and distinct from the holding in respect of which it was registered'.[33] Moreover, while not expressly deciding the issue, he was prepared to contemplate that the surrender of milk quota to the national reserve could constitute a part disposal of land. It would, therefore, be a matter of some interest to see the approach adopted by a court if required to address the tax treatment of the freehold sale of a holding together with all the quota attached.[34] Indeed, as shall be seen, in such circumstances Customs are prepared to accept that for the purposes of Value Added Tax there is no separate supply of quota.

In this context attention may also be redirected to the recently implemented derogation authorizing permanent transfers of quota without corresponding transfers of land.[35] Since the derogation marks a clear break in the link between milk quota and the holding, it may be interpreted as strong evidence in favour of the Revenue's argument that quota constitutes a distinct capital asset. Nevertheless, since permanent transfers without land are only available in certain carefully circumscribed circumstances in order to improve the structure of milk production,[36] it remains arguable that the tax treatment should likewise be an exception to the norm. Similar considerations would apply in the case of temporary transfers, again effected without the requirement of a land transaction. As a result it is possible to contemplate that, just as the milk quota legislation governing transfers adapts to meet different Community objectives, a similar degree of flexibility should be expected of the tax legislation.

2.3. Deemed disposals under section 22 of the Taxation of Chargeable Gains Tax Act 1992

Even if there is no actual disposal, there may yet be a deemed disposal within section 22 of the Taxation of Chargeable Gains Act 1992. As indicated, for such a deemed disposal to occur a capital sum must be derived

---

[33] Ibid., at 88.

[34] In addition, it has not been widely suggested that separate sums should be charged for rent of the land on the one hand and for use of milk quota on the other.

[35] Council Reg. (EEC) 3950/92, [1992] OJ L405/1, Art.8; and the Dairy Produce Quotas Regs. 1994, S.I.1994, No.672, reg.13, as amended by the Dairy Produce Quotas (Amendment) (No.2) Regulations 1994, S.I.1994 No.2919.

[36] By way of illustration, it has been provisionally calculated by MAFF that, of the 567 million litres permanently transferred during the 1994/5 milk year, a mere 11 million litres were transferred without land under the new provisions: *Third Report from the Agriculture Committee: Trading of Milk Quota* (Session 1994–5, H.C.512), at 47.

from an 'asset', a term broadly interpreted by the courts. For example, in *O'Brien* v. *Benson's Hosiery (Holdings) Ltd.* it was held that there could be a deemed disposal of the asset constituted by an employer's rights under a service contract, a key determinant being the ability to turn this contractual right to account.[37] Further, in *Zim Properties Ltd.* v. *Proctor* it was held that there could be a deemed disposal of the right to pursue a claim for professional negligence.[38] That having been said, it is clear that the mere payment of a capital sum does not in itself give rise to liability under section 22, since there must also be an asset from which the capital sum is derived. For example, Knox J in *Kirby* v. *Thorn EMI plc* did not accept that the right of a person to engage in trade amounted to an asset for the purposes of what is now section 22, notwithstanding that it could potentially be turned to account.[39] His reasoning was specifically adopted by the Special Commissioners in *Cottle* v. *Coldicott*, although they did not believe it applicable to milk quota. Whereas *Kirby* v. *Thorn EMI plc* concerned a liberty or freedom to trade enjoyed by everyone, the milk quota system had the effect of limiting participation in the market to producers with reference quantities.

In accordance with the principle that there must be an asset from which the capital sum is derived, the two most obvious candidates have been the milk quota, if correctly characterized as a separate asset, or the land or holding. In *Cottle* v. *Coldicott* it was held that the asset behind the capital payment was the quota; and that the quota in turn derived its value not from the holding but from the common organization of the market in milk or milk products.[40] Particular weight was placed on the decision of the European Court in *R.* v. *Ministry of Agriculture, Fisheries and Food*, ex parte *Bostock*, the Special Commissioners referring to the statement that the right to property does not include 'the right to dispose, for profit, of an advantage, such as the reference quantities allocated in the context of the common organization of a market, which does not derive from the assets or occupational activity of the person concerned'.[41] By contrast, the authority considered to most assist the appellant was *Pennine Raceway Ltd.* v. *Kirklees Metropolitan Council (No.2)*.[42] In that case the company enjoyed a licence from a landowner to conduct drag racing, planning permission having been obtained for that purpose. When the planning permission was revoked by the Council, the licence depreciated in value and compensation became payable. The compensation was held to be a capital sum derived from the licence and, accordingly, fell to be taxed under what is now section

---

[37] [1980] AC 562. [38] [1985] STC 90. [39] [1986] STC 200. [40] [1995] SpC 40.
[41] Case 2/92, [1994] ECR I–955, [1994] 3 CMLR 547, at I–984 (ECR) and 570 (CMLR). However, there is some ambiguity as as to whether the entity not derived from such assets or occupational activity is the right to dispose of quota for a profit or the quota itself.
[42] [1989] STC 122.

22(1)(a) of the Taxation of Chargeable Gains Act 1992.[43] However, the Special Commissioners considered that the decision was not on all fours with the case before them, since planning permission, as noted, cannot be transferred away from a particular plot of land.[44]

Since the common organization of the market was considered by the Special Commissioners to be the source of the value of milk quota, reference may also be made to *Davenport* v. *Chilver*.[45] In that case the Court accepted that a right to compensation conferred by statute could be an asset for the purposes of what is now section 22; and it may be argued that the value of milk quota is similarly derived from statute. Moreover, the case law reiterates that for the purposes of section 22 it is necessary to ascertain the real rather than the immediate source of the capital sum.[46] If this is the correct analysis, the taxpayer would be obliged to have recourse to ESC D33 for the purposes of securing, *inter alia*, roll-over relief. However, while there is the common emphasis on statute, the analogy would seem more appropriate with regard to the taxation of compensation payable under the Agriculture Act 1986.[47]

Before addressing the operation of the Capital Gains Tax legislation in certain specific situations, two further points of general application may be noted. First, if quota is a separate asset, it would now seem clear that taxpayers cannot rely on section 43 of the Taxation of Chargeable Gains Act 1992 so as to attribute to the quota a proportion of the allowable expenditure incurred on the land or holding.[48] In *Cottle* v. *Coldicott* the Special Commissioners rejected this contention, again on the basis that quota did not derive its value from the holding. Further, they were firmly of the view that in 1984 there had been no merger, division, or change of nature in the holding; nor had there been the creation or extinction of rights or interests in or over the holding.[49] Secondly, there is an argument that, if quota is to be treated as a separate asset, it is a wasting asset. Council

---

[43] The legislation in force at the operative date was the Capital Gains Tax Act 1979, s.20(1)(a).

[44] However, it may be noted that in *Pennine Raceway Ltd.* v. *Kirklees Metropolitan Council (No.2)* the capital sum was derived from the asset constituted by the licence rather than from the planning permission.

[45] [1983] STC 426.   [46] See, e.g., *Zim Properties Ltd.* v. *Proctor*, [1985] STC 90.

[47] The case of *Davenport* v. *Chilver* will be considered in greater detail in that context.

[48] The Taxation of Chargeable Gains Act 1992, s.43 provides that: 'If and in so far as, in a case where assets have been merged or divided or have changed their nature or rights or interests in or over assets have been created or extinguished, the value of an asset is derived from any other asset in the same ownership, an appropriate proportion of the sums allowable as a deduction in the computation of a gain in respect of the other asset under paragraphs (a) and (b) of section 38(1) shall, both for the purpose of the computation of a gain accruing on the disposal of the first-mentioned asset and, if the other asset remains in existence, on a disposal of that other asset, be attributed to the first-mentioned asset.' The provision in force as at the operative date in *Cottle* v. *Coldicott* was the Capital Gains Tax Act 1979, s.36.

[49] In this context, see, e.g., *Bayley* v. *Rogers*, [1980] STC 544.

Regulation (EEC) 3950/92 authorizes the continuation of the present system until the year 2000, but not beyond;[50] and on this basis quota's predictable life would not seem to exceed fifty years.[51]

2.4. Specific situations

The more detailed operation of the Capital Gains Tax legislation may be considered in four specific situations: first, with regard to permanent and temporary transfers; secondly, with regard to the availability of retirement relief; thirdly, with regard to compensation payable to departing tenants under the Agriculture Act 1986 or otherwise; and, fourthly, with regard to compensation for permanent cuts and under outgoers' schemes.

2.4.1. Permanent and temporary transfers

Proceeds from the permanent transfer of milk quota would seem to fall naturally under the Capital Gains Tax legislation;[52] and, as indicated, the main debate has centred upon whether there is an actual disposal of the quota, an actual disposal of an interest in or right over the land or holding, or some form of deemed disposal under section 22 of the Taxation of Chargeable Gains Act 1992. As has also been seen, statutory intervention has provided that, even if the Revenue is correct in the contention that quota constitutes a separate asset, it is none the less a 'qualifying asset' for the purposes of roll-over relief.[53] However, should the farmer making the disposal have entered into one or more temporary transfers, there is a possibility that entitlement to roll-over relief may be restricted on the basis that the quota ceases *pro tanto* to be used for a business.[54]

Proceeds from temporary transfers of milk quota are generally considered to be income in character; but there is an argument that such transactions could amount to a part disposal of the quota or the land or holding to which it is referable, by reason of the interaction of Schedule 8 to the Taxation of Chargeable Gains Act 1992 and section 34 of the Income and Corporation Taxes Act 1988. Under Schedule 8, where a lease of land or other property is granted for a premium, there will be a part disposal of the asset out of which the lease is granted. This Schedule may be sufficiently

---

[50] [1992] OJ L405/1, Art.1.

[51] The Taxation of Chargeable Gains Act 1992, s.44(1).

[52] In the *Tax Bulletin* of August 1994 the Revenue affirmed that, in the hands of a farmer who uses it for the business of milk production, quota is a fixed capital asset.

[53] See now the Taxation of Chargeable Gains Act 1992, s.155. If, by contrast, there is a deemed disposal analogous to that in *Davenport* v. *Chilver*, then it would seem necessary to place reliance on ESC D33.

[54] The taxpayer could advance the alternative view that the quota was being used for another (non-farming) business: see the Taxation of Chargeable Gains Act 1992, s.152(8).

broad to cover temporary transfers of quota.[55] Should that be so, temporary transfers would only fall outside the part disposal rules if charged to income tax under section 34 of the Income and Corporation Taxes 1988, the provision which covers, *inter alia*, rent. In this latter context the definition of 'lease' is somewhat more narrow, being confined to transactions involving land.[56] Accordingly, if the temporary transfer does not amount to a lease for the purposes of section 34, it should remain governed by the Capital Gains Tax legislation.

2.4.2. *Retirement relief*

There has been little dispute that, whether or not milk quota is a separate asset, it none the less qualifies as a 'business asset' capable of attracting retirement relief. That having been said, two areas of concern may be addressed.[57] First, the Revenue could challenge entitlement if the producer has effected numerous temporary transfers to other producers as opposed to employing the quota actively for milk production.[58] Secondly, there has been dispute as to the scale of the disposal necessary to attract retirement relief. In particular, taxpayers have argued that the sale of quota without a disposal of land may suffice on the basis that it amounts to the disposal of 'part of a business' within section 163(2) of the Taxation of Chargeable Gains Act 1992. This contention has not always been accepted by the Revenue.[59]

In cases concerning farmers the expression 'part of a business' has as a rule been restrictively interpreted. For example, in *McGregor* v. *Adcock* retirement relief was refused where the owner of a thirty-five acre farm sold 4.8 acres for development;[60] and Fox J held that in each case it was a question of fact 'whether there has been such an interference with the whole complex of activities and assets as can be said to amount to a disposal of the business or a part of a business'.[61] In the later decisions of *Atkinson* v. *Dancer* and *Mannion* v. *Johnston*, heard together, the taxpayers suffered equal lack of success, notwithstanding that the proportion of land sold was

---

[55] See the definition of 'lease' in the Taxation of Chargeable Gains Act 1992, Sch.8, para.10(1). In particular, the definition extends beyond leases of land.

[56] See the Income and Corporation Taxes Act 1988, s.24(1).

[57] On these aspects, see, e.g., Albrow, E., 'The Changing Face of Farming', (1990) 124 Taxation 572–3; and Williams, D. and Paskins, H., 'Milk Quota', (1993) 208 Tax Journal 13–15.

[58] See, e.g., (1994) Farm Tax Brief, Vol.9. No.2, 13.

[59] See, however, e.g., [1993] 30 EG 80; and (1993) Farm Tax Brief, Vol.8, No.8, 61–2. It may be noted that under the general rule a land transaction is required in any event to effect a permanent transfer of quota. That having been said, special considerations are raised by the application of the part disposal rules to the grant of a short-term tenancy at a rack rent: see the Taxation of Chargeable Gains Act 1992, s.37(1); and, e.g., Cardwell, M. N. and Lane, S., 'The Taxation of Milk Quota', [1994] BTR 501–26.

[60] [1977] 3 All ER 65.   [61] Ibid., at 69.

considerably greater.[62] Similarly, in *Pepper* v. *Daffurn* retirement relief was refused on the sale of a cattle yard with the benefit of planning permission.[63] The taxpayer had in fact disposed of the majority of his farm two years earlier (on which occasion retirement relief had been granted). At around the time of the earlier sale he had shifted the focus of his farming activities from rearing calves to grazing store cattle. The Court accepted the General Commissioners' finding that this switch amounted to a fundamental change in the farming enterprise. However, it was pointed out that the change had taken place some time before the sale of the cattle yard and, besides, had been prompted by the hope of realizing a development gain. Accordingly, on the basis of both timing and causation, retirement relief was refused. Although the case could be decided on these grounds, Jonathan Parker J also took the opportunity to affirm that proper weight must be placed upon the statutory words, as opposed to adopting the 'interference test' enunciated in *McGregor* v. *Adcock*.[64] However, the factors to be considered were left unspecified; and in any event he was not required to address the disposal of milk quota.

None the less, since the case of *Jarmin* v. *Rawlings* has provided not only comfort with regard to the timing of disposals but also guidance as to the effect of discontinuing dairy operations.[65] The taxpayer had for many years prior to October 1988 milked a dairy herd on his freehold land, which extended to some sixty-four acres. On this land were located a milking parlour and yard, a hay barn, implement shed, and cattle sheds. As at October 1988, the herd consisted of thirty-four animals and a full-time labourer was employed. The taxpayer also enjoyed the benefit of a milk quota. Being then sixty-one years old, he sold at auction the milking parlour and yard, the storage barn, and a small amount of riverside pasture, in total some 1.2 acres. Completion was fixed for 27 January 1989; and during the intervening period prior to completion fourteen cows were sold, the balance being transferred to another farm in the ownership of the taxpayer's wife. However, there was no sale of milk quota, it being 'informally transferred' to the wife's farm for the remainder of 1988/9 milk year. During subsequent milk years temporary transfers were effected in favour of third parties, the taxpayer engaging in the rearing and finishing of store cattle on the retained land. While the decision as such did not concern a disposal of milk quota, it did provide the opportunity to consider which of these various events could be taken into account and whether the events which could be taken into account constituted a disposal of part of a business. It was held

---

[62] [1988] STC 758. In *Mannion* v. *Johnston* the landholding of 78 acres was reduced by 35 acres in two tranches.

[63] [1993] STC 466. See also, e.g., Newth, J., 'Retiring too slowly', (1994) 132 Taxation 304–5.

[64] For similar reservations as to the 'interference test, see *Atkinson* v. *Dancer* and *Mannion* v. *Johnston*, [1988] STC 758.

[65] [1994] STC 1005.

to be highly artificial to look only at the contract date itself. Indeed, 'To ignore events after the making of the contract would be to ignore the performance of the disposal itself'. Accordingly, while due regard was given to dicta in *Atkinson* v. *Dancer* and *Mannion* v. *Johnston* which cautioned against treating separate disposals as one, Knox J accepted that on the facts account had to be taken of the sales of cattle between contract and completion, in that they were part of the same transaction as the disposal of the farmyard and milking shed. With that point decided, the transactions which could be taken into account were held to constitute the disposal of a part of a business.

Three aspects may be highlighted. First, in line with the approach taken in *Pepper* v. *Daffurn*, the taxpayer's dairy farming was considered to comprise a separate part of the farm business. Secondly, the taxpayer's case was not prejudiced by the retention of some cattle, the vast majority of the grazing land, and the milk quota, there being no requirement for all the assets used in the business to be sold. Rather, each case would depend upon its own facts. Finally, and critically, a business connoted activity, and on the facts the taxpayer had ceased the production and sale of milk as at completion. In this regard the sale of the milking parlour proved 'a vital ingredient', since it became possible to say that it 'caused such an interference with the whole complex of activities and assets as to amount to a disposal of part of a business, if that is the correct test'.[66]

Accordingly, it would seem relatively safe to assume that a shift from dairy farming to some other form of livestock or arable husbandry is capable of amounting to the disposal of a part of a business. That having been said, each case will be decided upon its own facts, with some scope for taking into account disposals on different days, provided that they form integral parts of the same transaction. However, it would be wrong to stretch this latitude too far, as recently illustrated in *Wase* v. *Bourke*.[67] The taxpayer sold his entire dairy herd in March 1988, but the sale of the quota was delayed until February 1989. The subsequent sale of the quota was held not to constitute the disposal of part of a business. Rather it was the disposal of a former business asset. The relevant business activity had been the production and sale of milk; and this had ceased on the disposal of the herd. Further, the interval between the two disposals was too great to satisfy the test as laid down in *Jarmin* v. *Rawlings*.

2.4.3. *Compensation payable to departing tenants*

Very difficult issues have arisen in connection with compensation payable to departing tenants; and these have been exacerbated by the fact that differ-

---

[66] Ibid., at 1015. Again, reference was made to the 'interference test' with reservations.
[67] *The Times*, 24 Nov. 1995.

ent factors seem to apply depending upon whether the sums are payable by virtue of the Agriculture Act 1986 or otherwise.[68] As indicated, if the various criteria of the Agriculture Act 1986 are satisfied, then the tenant enjoys a statutory entitlement to compensation under one or more of three heads, namely excess over standard quota, the tenants's fraction, and transferred quota. However, if the criteria are not satisfied, compensation is only payable to a tenant protected by the AHA 1986 where he has secured agreement to that effect with his landlord. In the case of farm business tenants under the ATA 1995, the statute provides for compensation, but this is dependent upon, *inter alia*, landlord's consent (or arbitrator's approval).

Much again rests upon the general considerations addressed earlier, not least whether milk quota is a separate asset or an interest in or right over the land or holding. Accordingly, alternatives would once more seem to be an actual disposal of the quota; an actual disposal of an interest in or right over the land or holding; or a deemed disposal under section 22 of the Taxation of Chargeable Gains Act 1992 (whether the asset from which the capital sum is derived be the quota, an interest in or right over the land or holding, or a statutory right). However, with regard to compensation payable under the Agriculture Act 1986 a fourth alternative may be suggested, namely that the sums received fall out of the tax net altogether by a analogy with statutory compensation payable under section 60 of the AHA 1986 or section 37 of the Landlord and Tenant Act 1954. In any event the Revenue has taken a firm view that all compensation paid to departing tenants is liable to Capital Gains Tax, while in practice apparently accepting that roll-over relief may be available.[69]

In this connection four aspects may be highlighted, all but the last having particular relevance to statutory compensation. First, in the case of statutory compensation it may be more difficult to maintain that there has been an actual or deemed disposal of an interest in or right over the land or holding. Although compensation is not payable unless there has been the termination of a tenancy, such an event is not in itself sufficient to sustain a claim. An illustration is provided by the circumstances of tenants who vacated before the Agriculture Act 1986 came into force on 25 September 1986, the European Court in *R. v. Ministry of Agriculture, Fisheries and Food*, ex parte *Bostock* conclusively deciding that they had no entitlement to compensation.[70]

[68] For compensation payable to departing tenants, generally, see Ch.5, 4.

[69] The Revenue's justification for their approach is not completely clear. See, however, correspondence between the Country Landowners Association and the Revenue in *Milk Quotas: M2/89* (Country Landowners Association, London, 1989), at 24–9. Whatever may be the precise justification, acceptance that roll-over relief is available can but be of assistance to departing tenants (the relief being particularly valuable in the case of payments for excess over standard quota, which would seem to lack any base cost).

[70] [1994] ECR I–955, [1994] 3 CMLR 547.

Secondly, the tax treatment of statutory compensation also seems to provide the best context for the argument that there is a deemed disposal of a statutory right. In particular, reference may again be made to the decision in *Davenport* v. *Chilver*.[71] By virtue of a statutory instrument the taxpayer had received a capital sum as compensation for Latvian property which had been confiscated by the Soviet Union in 1940. The Court held that the sums received were chargeable to Capital Gains Tax on the basis, *inter alia*, that the statutory right to compensation was itself a form of property and, therefore, an asset from which a capital sum could be derived. Once payment had been made by the Soviet Union to the Foreign Compensation Commission, the taxpayer enjoyed an independent proprietary right to share in the designated fund. If this analysis is also appropriate for statutory compensation in respect of milk quota the taxpayer's position would seem the more precarious. In particular, the Revenue would probably not accept that there was a base cost for the statutory right to compensation; and such an 'asset' would not qualify for roll-over relief under section 155 of the Taxation of Chargeable Gains Act 1992. Again there would remain the possibility of claiming relief under ESC D33; but it would need to be recognized that the grant of relief in such circumstances is concessionary.[72]

Thirdly, it would now seem more difficult to argue that statutory compensation escapes Capital Gains Tax altogether. Not only did this argument fail in *Davenport* v. *Chilver*,[73] but, more recently, in *Pennine Raceway Ltd.* v. *Kirklees Metropolitan Council (No.2)* Croom-Johnson LJ expressly rejected any 'general proposition' that compensation awarded by statute falls outside what is now section 22 of the Taxation of Chargeable Gains Act 1992.[74] That having been said, no tax was held payable in *Davis* v. *Powell*[75] and *Drummond* v. *Brown*,[76] and both cases were directly concerned with the payment of compensation on the termination of tenancies, the former

---

[71] [1983] STC 426.

[72] The decision in *Davenport* v. *Chilver* may be contrasted with that in *Pennine Raceways Ltd.* v. *Kirklees Metropolitan Council (No.2)*, [1989] STC 122. In the latter case, as has been seen, the taxpayer enjoyed the benefit of a licence granted by a landowner to conduct drag racing, planning permission having been obtained for that purpose. On the revocation of the planning permission, compensation was received under the Town and Country Planning Act 1971, s.164(1). The Court of Appeal held that the compensation was indeed chargeable as a capital sum derived from an asset. However, the asset in question was identified as the licence rather than the statute, the licence being the real (rather than the immediate) source (following the test set out in *Zim Properties Ltd.* v. *Proctor*, [1985] STC 90). Likewise, should quota be treated as a separate asset, there is an argument that it too constitutes the real source of the compensation.

[73] [1983] STC 426.					[74] [1989] STC 122, at 130.

[75] [1977] STC 32 (in respect of compensation for disturbance payable under what is now the AHA 1986, s.60.

[76] [1984] STC 321 (in respect of compensation payable under the Landlord and Tenant Act 1954, s.37).

specifically relating to the agricultural sector.[77] Further, those decisions had emphasized that in reality the departing tenant was being reimbursed for his costs; and statutory compensation for milk quota is directed to ensuring that the tenant is not deprived of the fruits of his labours and capital outlay.[78]

In this context it may be emphasized that the need to address the tax treatment of statutory compensation under the Agriculture Act 1986 is likely to diminish over time. Fewer and fewer tenants will be able to satisfy the criterion that they were allocated quota or (in the case of transferred quota) that they were in occupation as at 2 April 1984;[79] and, further, the ATA 1995 expressly precludes entitlement under the Agriculture Act 1986 in the case of farm business tenancies.[80] As indicated, tenants who lack a statutory right to compensation under the Agriculture Act 1986 are placed at a disadvantage. Those whose tenancies remain governed by the AHA 1986 will be dependent upon securing their landlord's agreement; and, while those within the farm business tenancy régime do enjoy the benefit of statutory compensation,[81] a prerequisite to payment is landlord's consent to the improvement or, in the event of dispute, arbitrator's approval.

Fourthly, where compensation is payable purely by agreement, there are grounds for believing that a capital sum is derived from a contractual right.[82] Difficult questions would then arise as to the base cost of this contractual right. It would seem to be in the interest of the taxpayer to argue that the right came into being by way of a bargain at arm's length, so as to achieve a base cost equal to market value.[83] In the light of *Zim Properties Ltd.* v. *Proctor* this argument may be hard to sustain.[84] At the same time, the general prerequisite of landlord's consent under the 1995 Act raises just a possibility that the statute is not the only source of the payment.[85]

---

[77] The same approach to compensation under s.60 of the AHA 1986 was adopted by the Special Commissioners in the two recent cases of *Davis* v. *Henderson* and *Pritchard* v. *Purves*, respectively [1995] SpC 46 and 47. The compensation was held to be free from Capital Gains Tax; but emphasis was laid on the need for the tenant to vacate in consequence of the notice to quit.

[78] For the Revenue's reaction to the argument that no tax is payable, see the correspondence between the Country Landowners Association and the Revenue in *Milk Quotas: M2/89*. However, no mention is made of *Davis* v. *Powell*.

[79] This general rule is subject to the exceptions applicable in the case of certain successors, assignees, and head-tenants.

[80] The ATA 1995, s.16(3).

[81] Moreover, it is not possible to contract out of the compensation provisions: ibid., s.26.

[82] See, e.g., *O'Brien* v. *Benson's Hosiery (Holdings) Ltd.*, [1980] AC 562. However, it is not impossible that the quota would be seen as the real source of the capital sum: *Pennine Raceway Ltd.* v. *Kirklees Metropolitan Council (No.2)*, [1989] STC 122.

[83] The Taxation of Chargeable Gains Act 1992, s.17.   [84] [1985] STC 90.

[85] For a fuller discussion of the 1995 Act, see De Souza, J., 'Compensation for Agricultural Improvements: a most Unsatisfactory Tangle', together with correspondent's comments, (1995) 2 Private Client Business 328–31.

2.4.4. Compensation for permanent cuts and under outgoers' schemes

The Revenue's current position is that compensation for permanent cuts should be treated as a capital sum derived from quota, giving rise to a deemed disposal under section 22(1)(a) and (2) of the Taxation of Chargeable Gains Act 1992.[86] Prior to 1991 compensation for cuts was taxable in the first year of receipt, but as from 1991 such payments have been treated as chargeable in the years or years of actual receipt. This change in practice has been of particular benefit to producers who opted to discontinue milk production definitively in return for payment by annual instalments over the period 1992–6. In their case each payment is treated as a separate disposal, so allowing maximum advantage to be taken of the annual exemption.[87] That having been said, it may again be questioned whether the Revenue is correct in its assertion that there is a deemed disposal of quota. Not least, in *Faulks* v. *Faulks* Chadwick J expressly contemplated that the surrender of quota to the reserve could amount to a part disposal of the land.[88]

3. INCOME TAX

Assuming that temporary transfers give rise to a charge to Income Tax (as opposed to a part disposal under the Capital Gains Tax legislation), it remains to ascertain whether the proceeds fall within Schedule A, Schedule D, Case I, or Schedule D, Case VI.

Schedule A covers the receipts of a person which arise 'from, or by virtue of, his ownership of an estate or interest in or rights over' land in the United Kingdom.[89] Accordingly, if quota is an interest in land, Schedule A would seem to apply. Again *Faulks* v. *Faulks* provides perhaps the strongest support for this proposition; but the strength of the support is somewhat weakened by the characterization of temporary transfers as transfers of 'potential liability'.[90] Moreover, taxation under Schedule A sits uneasily with the fact that such transactions are exceptions to the general principle under which quota is linked to the holding.

Should Schedule A not apply, a further consideration is whether receipts from temporary transfers qualify as trade receipts. If they do not, Schedule D, Case VI would seem to apply — and there is evidence that the Revenue

[86] *Tax Bulletin* of May 1994.

[87] Ibid.; and for the Community legislation, see Council Reg. (EEC) 1630/91, [1991] OJ L150/19; and Council Reg. (EEC) 1637/91, [1991] OJ L150/30. See also *MAFF News Release* 261/91.

[88] [1992] 15 EG 82.　　　　　[89] The Income and Corporation Taxes Act 1988, s.15(1)(c).

[90] [1992] 15 EG 82, at 88.

has adopted such an interpretation.[91] However, even if they can be demonstrated to be trade receipts and taxable under Schedule D, Case I, it is not necessarily clear whether 'quota leasing' will constitute an integral part of the farming business or a separate business. Differing sets of facts may give rise to differing results. For example, an active producer who effects a temporary transfer of a small proportion of his reference quantity every few years would have reasonable grounds for maintaining that the transactions form part of his dairy business. By contrast, a non-active quota holder who annually enters into temporary transfers of his entire reference quantity would seem unlikely to enjoy the same success.[92]

Finally, the Revenue would appear to treat compensation payable in respect of temporary cuts as income, and to allow sums paid by way of superlevy as a Schedule D, Case I deduction.[93]

4. INHERITANCE TAX

By contrast, the régime governing Inheritance Tax has been substantially more benign. The Revenue's *Tax Bulletin* of February 1993 contained an unequivocal statement that 'where agricultural land, or an interest in agricultural land, is valued and the valuation of the land reflects the benefit of milk quota, agricultural relief is given on that value'. However, if there is a separate valuation of the quota, then 'it will normally constitute an asset used in the business' within section 110 of the Inheritance Tax Act 1984; and, as a result, business property relief may be available.

In essence, agricultural property relief or business property relief would seem to be available dependent upon the method of valuation. That having been said, since the qualifying rules for the respective reliefs are slightly different, it would be to the taxpayer's advantage to take full note of those qualifying rules before choosing to value the quota with the land or independently. None the less, the more generous approach largely precludes the need to ascertain whether milk quota is a separate asset, while recognizing that its role in the farming business is such as to deserve one form of relief or the other.

---

[91] See, e.g., (1994) Farm Tax Brief, Vol.9, No.2, 13. The Inspector concerned took the view that the quota had acquired the character of an 'investment asset'.

[92] Since 1 Apr. 1994 it has been possible for producers to effect temporary transfers of their entire reference quantity: the Dairy Produce Quotas Regs. 1994, S.I.1994 No.672, reg.15. Even before that date the Scottish Land Court held that obtaining an income from temporary transfers of quota was not 'a farming operation': *Cambusmore Estate Trusteees* v. *Little*, [1991] SLT (Land Ct.) 33.

[93] For the ability to deduct superlevy payments, see *Tax Bulletin* of August 1994.

Where the qualifying rules are satisfied, the benefit of these reliefs is not easy to overestimate. The full rate is 100 per cent in the case of business property;[94] and for agricultural property the full rate is again 100 per cent (the requisite condition in this latter case being that the transferor enjoys vacant possession or the right to it within twelve months, or, following recent amendment, that the property is let on a tenancy beginning on or after 1 September 1995).[95]

5. VALUE ADDED TAX

With regard to Value Added Tax much again would seem to depend upon whether quota is rightly characterized as a separate asset. If it is, then the supply of quota should be subject to Value Added Tax at the standard rate, as neither exemption nor zero-rating appear applicable. If, by contrast, quota is to be treated as an interest in or right over land, then any supply of that interest or right should be exempt — unless the person making the supply has exercised the option to tax.[96]

For practical purposes considerable assistance can be obtained from Customs' *Business Brief 17/94*. It would appear that the approach adopted in that document is to determine whether the main purpose of the transaction is to make a supply of land or to make a supply of quota. Accordingly, where there is a permanent transfer of quota without land for the purposes of restructuring, as authorized by Article 8 of Council Regulation (EEC) 3950/92,[97] in Customs' view the transaction is a supply of services and chargeable at the standard rate. Although not expressly stated, the same treatment would presumably be considered applicable in the case of temporary transfers. By contrast, in the event of a permanent transfer of quota with land under one agreement, Customs are of the opinion that there is a single supply of land; and that the supply is exempt unless the option to tax has been exercised. Moreover, the same treatment is to be applied where the consideration is split between the land and the quota.

This approach to permanent transfers with land would seem consistent with *Faulks* v. *Faulks*;[98] and also with the Value Added Tax principle that there is only one supply where a transaction has two elements but one is

---

[94] The Finance (No.2) Act 1992, Sch.14.

[95] Ibid.; and the Finance Act 1995, s.155.

[96] In this context see again, *Cottle* v. *Coldicott*, [1995] SpC 40.

[97] [1992] OJ L405/1. For the UK legislation, see the Dairy Produce Quotas Regs. 1994, S.I.1994 No.672, reg.13, as amended by the Dairy Produce Quotas (Amendment) (No.2) Regs. 1994, S.I.1994 No.2919.

[98] [1992] 15 EG 82.

'incidental to' or 'integral to' the another.[99] Accordingly, notwithstanding that quota may be a separate asset, if its provision is 'incidental to' or 'integral to' an exempt supply of land (on the assumption that the option to tax has not been exercised), then the quota too may be transferred without incurring a charge to tax.

However, Customs adapt this general analysis in the case of permanent transfers through the medium of short-term agreements. In *Business Brief 17/94* the stance adopted is that, where milk quota is 'transferred with a grazing licence', there are two separate supplies. One is a supply of animal feeding-stuffs through the grazing licence, which is zero-rated; and the other a supply of quota, which is standard-rated. As a preliminary point it could be argued that less than full account is taken of the current transfer provisions, since it has not been possible to transfer quota through the medium of a grazing licence since 1 April 1988.[100] If, however, the same interpretation were applied to transfers effected through the medium of short-term tenancies, then there would seem to be an exempt supply of the land (unless the option to tax has been exercised) and a standard-rated supply of quota. Support could be derived from the decision of the Special Commissioners in *Cottle* v. *Coldicott*, which (for Capital Gains Tax purposes) expressly considered transfers through the medium of short-term tenancies and likewise identified quota as a separate asset. Indeed, there may be benefit in accepting Customs' interpretation, since it is not impossible to argue that the tenancy provides no more than a vehicle for the transfer of quota in compliance with Community and national rules and, as a result, is 'incidental to' or 'integral to' the supply of quota. On that basis there would be a single standard-rated supply of quota.

Business Brief 17/94 does not cover the Value Added Tax treatment of compensation in respect of milk quota payable to departing tenants. However, compensation under the AHA 1986 had been addressed somewhat earlier in Notice 742B of 1 January 1990.[101] Where paid following a notice to quit in accordance with the statutory procedure, compensation was stated to fall outside the scope of the tax. On the other hand, sums paid over and above the statutory amount were regarded as consideration for a standard-rated supply of services, as were sums paid following voluntary negotiations. There was no comment whether the same treatment would be applied to compensation under the Agriculture Act 1986. That having been said, the dangers faced by a tenant have now been substantially reduced by the

---

[99] See, e.g., *British Airways plc* v. *Customs and Excise Commissioners*, [1990] STC 643; *Customs and Excise Commissioners* v. *United Biscuits (UK) Ltd.*, [1992] STC 325; and *Rayner and Keeler Ltd.* v. *Customs and Excise Commissioners*, [1994] STC 724.

[100] The Dairy Produce Quotas (Amendment) Regs. 1988, S.I.1988 No.534, reg.5.

[101] At para.13.

decision of the European Court in *Lubbock Fine & Co.* v. *Customs and Excise Commissioners.*[102] In that decision the consideration paid for the surrender of a lease was held to be exempt from Value Added Tax where the rent was exempt.[103]

[102] Case 63/92, [1993] ECR I–6665, [1994] 3 WLR 261, [1994] 2 CMLR 633.
[103] In *Lubbock Fine* the consideration had been paid by agreement.

7

Conclusion

More than a decade having elapsed since the introduction of milk quotas, it may now be possible to hazard certain more general conclusions. Three aspects, in particular, may be addressed: the extent to which the quota system has met its stated objectives; the operation of the transfer provisions; and the adequacy of protection for tenants.

From inception it has been the twin aim of the milk quota system to curb the increase in milk production while at the same time permitting the necessary structural developments and adjustments. As indicated, the former aim has met with some success. With the passage of time it is easy to forget that at the time of the Commission's proposals to introduce quotas, the dairy sector was adjudged to be different from other agricultural sectors 'by virtue of the unremitting and even accelerating divergence of the trends of production and consumption'.[1] Moreover, the very future of the Common Agricultural Policy was considered to be in jeopardy.[2] Such fears would now seem to have been allayed. Consumption may have proved difficult to encourage, but production has been significantly reduced, notwithstanding the Nallet Package and SLOM allocations. Thus, the United Kingdom guaranteed total quantity for wholesale quota fell from 15,487,000 tonnes in the 1984/5 milk year[3] (together with a further 65,000 tonnes for Northern Ireland) to 14,197,179,000 tonnes in the 1993/4 milk year.[4] Extending the comparison to the ten Member States, 1983 deliveries were some 10 per cent higher than the combined guaranteed total quantities for wholesale quota in the 1994/5 milk year.[5] Since 1992 stocks of butter and skimmed-milk powder are also substantially reduced.[6] In addition, the dairy sector has ceased to dominate the agricultural budget. For example, in the 1995 budget it accounted for little more than 10 per cent of European Agricultural Guidance and Guarantee Fund Guarantee expendi-

---

[1] COM(83)500, at 15. [2] Council Reg. (EEC) 856/84, [1984] OJ L90/10, Preamble.
[3] Council Reg. (EEC) 1557/84, [1984] OJ L150/6.
[4] Council Reg. (EEC) 1560/93, [1993] OJ L154/30. For subsequent adjustment to take account of conversions between wholesale and direct sales quota, see Council Reg. (EC) 647/94, [1994] OJ L80/16.
[5] Residuary Milk Marketing Board, *EC Dairy Facts and Figures: 1994 Edition*, at Table 21.
[6] See, e.g., Residuary Milk Marketing Board, *EC Dairy Facts and Figures: 1994 Edition*, at Tables 65 and 82; and European Commission, *The Agricultural Situation in the Community: 1994 Report*, at T/291.

ture;[7] and at the same time such Guarantee expenditure now forms a smaller share of the total Community budget.[8] That having been said, persistent difficulties remain, not least high butterfat levels (which have consistently triggered the superlevy in the United Kingdom); and the entrenched approach of a quota system may not be the most appropriate for exploiting emerging world markets. Thus, the Community share of world trade in butter has fallen from 59.4 per cent in 1987 to 24.5 per cent in 1993.[9] At the same time, as highlighted by the reform of the milk marketing régime, the United Kingdom is effectively locked into a system under which it cannot advance to achieve national self-sufficiency in all forms of milk product.[10]

With regard to structural development, there would seem to be compelling evidence that milk quotas have not caused the ossification of the dairy industry.[11] The trend towards fewer and fewer producers operating increasingly efficient herds would seem to continue inexorably in the absence of national measures to the contrary; and it may be reiterated that in the case of the United Kingdom the number of registered producers has fallen from 50,625 in March 1984 to 36,709 in March 1994.[12] Moreover, the high prices now commanded by quota can but increase the barriers faced by prospective new entrants. As indicated, the House of Commons Agriculture Committee has expressed concern at their predicament, without feeling able to recommend immediate remedial measures.[13] This state of affairs may be sharply contrasted with the recent reform of agricultural tenancies effected under the ATA 1995, an avowed purpose of which is to increase the supply of tenancies for young farmers. In the dairy sector there must be a danger that such purpose will in part be frustrated by the costs of acquiring milk quota, whether by permanent or temporary transfer. At the same time this is a suitable context for reaching a considered assessment of the importance to be attached to the social dimension of the Common Agricultural Policy.[14]

The operation of the transfer provisions has without doubt proved central to the operation of the milk quota system as a whole. In this light the

[7] The European Commission, *The Agricultural Situation in the Community: 1994 Report*, at T/90.

[8] Ibid., at 116.   [9] Ibid., at T/285.

[10] In effect, national self-sufficiency has ceded to Community self-sufficiency.

[11] On this aspect, generally, see Ch.3, 1.3.

[12] England and Wales Residuary Milk Marketing Board, *United Kingdom Dairy Facts and Figures: 1994 Edition*, at Table 1.

[13] *Third Report from the Agriculture Committee: Trading of Milk Quota* (Session 1994–5, H.C.512), at xv. A further Report is to be completed.

[14] It may be useful to contrast the relative weight attached to the social dimension in respectively the Dairy Produce Quotas Regs. 1994 and the milk quota regulations recently adopted by Ireland, the European Communities (Milk Quota) Regs. 1995, S.I.1995 No.266.

extent of the legal uncertainty which surrounds them may be considered unfortunate — but perhaps an inevitable product of Member States implementing Community rules so as to best accomodate national needs. However, the general principle that reference quantities are linked to holdings has consistently been defended by the Community institutions, not least to prevent quota becoming the subject of speculation.[15] In the United Kingdom a similar approach was adopted in *Faulks* v. *Faulks*.[16] That having been said, it is also established that derogations from the general principle are to be permitted where they constitute a benefit to the system. Accordingly, since the Dairy Produce Quotas Regulations 1994 came into force on 1 April 1994, producers in the United Kingdom have been able to effect temporary transfers of their entire reference quantities without the requirement of a land transaction; and they can also effect permanent transfers without land in order to improve the structure of milk production.[17] The greatest difficulties would rather appear to be raised by permanent transfers through the medium of short-term tenancies. Although under the current legislation the grant of a tenancy of ten months or more is sufficient to trigger a transfer of quota in England and Wales,[18] such arrangements should still be treated with circumspection. For example, in *R.* v. *Ministry of Agriculture, Fisheries and Food*, ex parte *Cox* the Court made clear that some physical use of the land was required;[19] and in *Cottle* v. *Coldicott* the Special Commissioners held that quota transferred by this mechanism was to be regarded as a separate asset for Capital Gains Tax purposes.[20] Indeed, the Court of Auditors has construed such transactions as transfers 'without land',[21] the primary objective, namely the transfer of quota from the holding, being emphasized at the expense of the land transaction.

For the time being Government policy is to retain flexibility in the operation of the quota transfer provisions. Such flexibility has been advocated on the basis that it loosens the economic restraint imposed by the quota system and gives rise to an efficient industry.[22] Not least, with producers offered a

---

[15] See, e.g., Council Reg. (EEC) 3950/92, [1992] OJ L405/1, Preamble; the written observations of the Commission in Case 5/88, *Wachauf* v. *Bundesamt für Ernährung und Forstwirtschaft*, [1989] ECR 2609, [1991] 1 CMLR 328; and Case 98/91, *Herbrink* v. *Minister van Landbouw, Natuurbeheer en Visserij*, [1994] ECR I–223, [1994] 3 CMLR 645.

[16] [1992] 15 EG 82.

[17] S.I.1994 No.672, regs.13 and 15, as now amended by the Dairy Produce Quotas (Amendment) (No.2) Regs. 1994, S.I.1994 No.2919. For the Community legislation, see Council Reg. (EEC) 3950/92, [1992] OJ L405/1, Arts.6 and 8.

[18] The Dairy Produce Quotas Regs. 1994, S.I.1994 No.672, reg.7(6)(a)(ii).

[19] [1993] 22 EG 111, [1993] 2 CMLR 917.   [20] [1995] SpC 40.

[21] *Special Report No.4/93 on the Implementation of the Quota System intended to control Milk Production*, [1994] OJ C12/1, at 4.43.

[22] See, e.g., *Third Report from the Agriculture Committee: Trading of Milk Quota* (Session 1994–5, H.C.512), at xiii.

ready opportunity to turn unused reference quantities to account, there is
liable to be fuller use of the national quota. This commercially orientated
approach has also found expression in calls for cross-border transfers,[23] it
being apprehended that quota will migrate to those northern regions where
the climate is best suited to milk production.[24] As yet there is no certainty
that these calls will lead to material amendment.[25] Much may depend upon
the priority given to the overall competitiveness of the Community dairy
industry as against national and regional interests.

Another aspect of the transfer provisions which awaits clear resolution
is that of business reorganizations. In large part this may be attributed to
the diversity of business structures throughout the various Member
States; and in *Von Deetzen II* the European Court stated unequivocally
that detailed consideration of specific cases would be a matter for national
courts.[26] In the United Kingdom difficult issues are raised, for example,
by changes in the composition of partnerships. As has been seen,
such changes can occur without a land transaction sufficient to trigger a
transfer of quota under the Dairy Produce Quotas Regulations 1994.[27]
That having been said, there is Court of Appeal authority to the effect
that the quota may yet be transferred;[28] and reference may again be made
to *Von Deetzen II*, where greater weight was accorded to the commercial
reality of the business reorganization than to its precise form. Such a degree
of uncertainty can hardly assist producers. However, in view of the complex
nature of quota and the variety of business structures, it is difficult to
the escape the conclusion reached by the European Court that much will
indeed depend upon the detailed facts of the immediate case before the
national court.

With regard to landlord and tenant relations, the Agriculture Act 1986
may be seen as a typical of the close regulation which prevailed in the
agricultural sector prior to the ATA 1995, with the extent of specific provision perhaps best illustrated by the alternative methods of calculating 'excess over standard quota'. None the less, the relative absence of litigation
would suggest that generally it struck a reasonable balance between landlord and tenant. Further, such an interpretation was accepted before the
European Court in *R.* v. *Ministry of Agriculture, Fisheries and Food*, ex

[23] See, e.g., *MAFF News Release* 68/94.
[24] Proper consideration of this facility has been advocated by the Court of Auditors in their
*Special Report No.4/93 on the Implementation of the Quota System intended to control Milk
Production*, [1994] OJ C12/1, at 5.17.
[25] It may be noted that as recently as 1993 a 'holding' became expressly restricted to the
territory of a single Member State: Council Reg. (EEC) 1560/93, [1993] OJ 154/30, amending
Council Reg. (EEC) 3950/92, [1992] OJ L405/1, Art.9(d).
[26] [1991] ECR I–5119, [1994] 2 CMLR 487.   [27] See, generally, Ch.4, 4.3.
[28] *W. E. & R. A. Holdcroft* v. *Staffordshire County Council*, [1994] 28 EG 131.

parte *Bostock*.[29] The farm business tenancy régime also provides for statutory compensation; but in general such compensation is dependent upon securing landlord's written consent. As seen, while there is the facility to refer the matter to arbitration, the requisite notice must be served before the improvement is commenced. Accordingly, if a tenant without written consent begins to attach quota to any part of the tenanted land, he will forfeit the opportunity to go to arbitration. Although in theory there remains the possibility of agreeing terms with his landlord, in practice he will have severely prejudiced his negotiating position. Moreover, in the light of the demand for farm business tenancies, it will be of interest to note the approach adopted by landlords towards tenants who seek prior written consent.

Finally, under the current legislation the milk quota system is to continue until the year 2000.[30] Although there has been a history of extensions following the initial five year period, further extensions cannot be guaranteed.[31] The United Kingdom Government openly opposes the retention of the current régime. In particular, the Minister of Agriculture has adopted the arguments against quotas set out in *European Agriculture: the Case for Radical Reform*.[32] Driving forces behind change are perceived to be GATT commitments under the Uruguay Round and prospective enlargement of the European Union; and the response advocated is a more market-orientated policy which promotes competitiveness and efficiency. It may also be observed that the publication expressly rejects the concept of a two-tier price structure, under which a specified amount of production would benefit from support but the excess would be exposed to world market forces. Accordingly, there is little sympathy with the concept of 'B-Quotas' in the form of unsupported production in excess of individual reference quantities.[33] This more market-orientated approach could only accelerate the structural trends towards larger dairy enterprises; and it remains to be seen whether this is politically desirable, or indeed acceptable, throughout the Member States. At the same time significant change cannot be ruled out even prior to the year 2000. As the case law of the European Court has

---

[29] Case 2/92, [1994] ECR I–955, [1994] 3 CMLR 547. Nevertheless, in certain circumstances the Act could work very much to the detriment of the landlord or tenant: see, e.g., *Creear* v. *Fearon*, [1994] 46 EG 202.

[30] Council Reg. (EEC) 3950/92, [1992] OJ L405/1, Art.1.

[31] Indeed, prognosis as to the survival of quotas is understood to be a factor underlying their market value. Should quotas not survive beyond the year 2000, of some importance will be the extent to which their value is translated back into the land.

[32] *MAFF News Release* 285/95. The publication represented the conclusions of MAFF's CAP Review Group (which comprised 11 independent experts).

[33] For the time being Community progress towards the implementation of B-Quotas would appear unlikely: see, e.g., (1994) Dairy Industry Newsletter, Vol.6, No.14, 4.

made plain, producers cannot legitimately expect that existing situations within the ambit of the Community institutions' discretionary power will continue unaltered. As a result they cannot claim a vested right to the maintenance of an advantage obtained from a common organization of the market.[34]

Whatever the future may bring, there can be no denying that milk quotas have made a major contribution to Community jurisprudence. Not least, they constitute the complex and comprehensive regulation of an important sector of economic activity. The degree of this regulation is perhaps the more remarkable in the light of general advocacy of free trade in the last decade of the twentieth century. Consequently, the milk quota system continues to provide a test bed for exploring the extent to which legislation may achieve curbs on production in tandem with progress towards economic efficiency.

---

[34] See, e.g., Case 230/78, *S.p.A. Eridania-Zuccherifici nazionali and S.p.A. Italiana per l'industria degli Zuccheri* v. *Minister of Agriculture*, [1979] ECR 2749; Case 52/81, *Faust* v. *Commission*, [1982] ECR 3745; and, more recently, Case 280/93, *Germany* v. *Council*, [1994] ECR I–4973.

Further Reading

N.B. This list of further reading largely excludes more general works on Community law, administrative law, and the law of landlord and tenant.

BOOKS

Ackrill, R., *Information Sources on the Common Agricultural Policy* (European Information Association, 1994).

Anderson, H., *Agricultural Charges and Receivership* (Chancery Law Publishing, London, 1992).

Apsion, G., *Milk Quotas* (Farm Tax and Finance Publications, 1992), together with *1993 Supplement*.

Barents, R., *The Agricultural Law of the EC* (Kluwer, Deventer, 1994).

Brown, L. N. and Kennedy, T., *The Court of Justice of the European Communities* (4th edn., Sweet and Maxwell, London, 1994).

Burrell, A. M. (ed.), *Milk Quotas in the European Community* (CAB International, Wallingford, Oxfordshire, 1989).

Densham, H. A. C., *Scammell and Densham's Law of Agricultural Holdings* (7th edn., Butterworths, London, 1989), together with *1993 Supplement*.

Dillen. M. and Tollens, E., *Milk Quotas: their Effects on Agriculture in the European Community* (Eurostat, 1990).

Evans, D., *The Agricultural Tenancies Act 1995* (Sweet and Maxwell, London, 1995).

Fennell, R., *The Common Agricultural Policy of the European Community* (2nd edn., BSP Professional Books, Oxford, 1987).

Gehrke, H., *The Implementation of the EC Milk Quota Regulations in British, French and German Law* (European University Institute, Florence, 1993).

Gregory, M. and Sydenham, A., *Essential Law for Landowners and Farmers* (3rd edn., Blackwell, Oxford, 1990).

Harris, S., Swinbank, A., and Wilkinson, G., *The Food and Farm Policies of the European Community* (John Wiley, Chichester, 1983).

Harvey, D. R., *Milk Quotas: Freedom or Serfdom?* (Centre for Agricultural Strategy, 1985).

Hubbard, L. J., *Some Estimates of the Price of Milk Quota in England and Wales* (Department of Agricultural Economics and Food Marketing, University of Newcastle-upon-Tyne, 1991).

Lennon, A. A. and Mackay, R. E. O. (edd.), *Agricultural Law, Tax and Finance* (Looseleaf, Longmans, London).

McInerney, J. P. and Hollingham, M. A., *Readjustments in Dairying: an Analysis of Changes in Dairy Farming in England and Wales following the Introduction of Milk Quotas* (Agricultural Economics Unit, University of Exeter, 1989).

Milk Marketing Board, *Five Years of Milk Quotas: a Progress Report* (Thames Ditton, Surrey, 1989).

Moody, J. (with Jessel, C.), *Agricultural Tenancies Act 1995 — a Practical Guide* (Farrer and Co., London, 1995).

Muir Watt, J., *Agricultural Holdings* (13th edn., Sweet and Maxwell, London, 1987), together with *1989 Supplement*.

Neville, W. and Mordaunt, F., *A Guide to the Reformed Common Agricultural Policy* (Estates Gazette, London, 1993).

Neville-Rolfe, E., *The Politics of Agriculture in the European Community* (Policy Studies Institute, London, 1984).

Oskam, A. J., van der Stelt-Scheele, D. D., Peerlings, J., and Strijker, D., *The Superlevy — is there an Alternative?* (Wissenschaftsverlag Vauk Kiel, 1988).

Petit, M., de Benedictis, M., Britton, D., de Groot, M., Henrichsmeyer, W., and Lechi, F., *Agricultural Policy Formation in the European Community: the Birth of Milk Quotas and CAP Reform* (Elsevier, Amsterdam, 1987).

Rodgers, C. P., *Agricultural Law* (Butterworths, London, 1991).

Bates, St. J., Finnie, W., Usher, J. A., and Wildberg, H. (edd.), *In Memoriam J. D. B. Mitchell* (Sweet and Maxwell, London, 1983).

Schermers, H. G., Heukels, T., and Mead, P., edd., *Non-contractual Liability of the European Communities* (Europa Instituut, University of Leiden, Nijhoff, 1988).

Slatter, M. and Barr, W., *Farm Tenancies* (BSP Professional Books, Oxford, 1987).

Snyder, F. G., *Law of the Common Agricultural Policy* (Sweet and Maxwell, London, 1985).

Stanley, O., *Taxation of Landowners and Farmers* (Looseleaf edn., Butterworths, London, 1995).

Stratton, R., Sydenham, A., and Baird, A., *Share Farming* (3rd edn., CLA Publications, London, 1992).

Sydenham, A. and Mainwaring, N., *Farm Business Tenancies: Agricultural Tenancies Act 1995* (Jordans, Bristol, 1995).

Thomson, K. J. and Warren, R. M. (edd.), *Price and Market Policies in European Agriculture* (University of Newcastle-upon-Tyne, 1984).

Usher, J. A., *Legal Aspects of Agriculture in the European Community* (Clarendon Press, Oxford, 1988).

Wood, D., Priday, C., Moss, J. R., and Carter, D., *Milk Quotas: Law and Practice* (Farmgate Communications, 1986).

Wood, D., Priday, C., Moss, J. R., and Carter, D., *The Handbook of Milk Quota Compensation* (Farmgate Communications, 1987).

ARTICLES IN JOURNALS

Sir Crispin Agnew of Locknaw Bt, 'Apportionment of Milk Quota', [1992] Journal of the Law Society of Scotland 29–32.

Albrow, E., 'The Changing Face of Farming', (1990) Taxation 572–3.

Avery, G., 'The Common Agricultural Policy: a Turning Point?', (1984) 21 C.M.L.Rev. 481–504.

Cardwell, M. N., 'General Principles of Community Law and Milk Quotas', (1992) 29 C.M.L.Rev. 723–47.

Cardwell, M. N. and Lane, S., 'The Taxation of Milk Quota', [1994] B.T.R. 501–26.

Craig, P. P., 'Legitimate Expectations: a Conceptual Analysis', (1992) 108 L.Q.R. 79–98.

Fitzpatrick, F., 'The Legal Nature of Milk Quota', (1992) 185 Tax Journal 4–5.

Fitzpatrick, F., 'Milk Quota: a Hybrid Animal', (1993) 207 Tax Journal 12–13.

Harris, D., 'Sorting the Sheep from the Cows', (1993) 221 Tax Journal 14–16.

Heukels, T., *Mulder II*, (1993) 30 C.M.L.Rev. 368–86.

Hill, S. O., 'Milk Quotas: Legal and Taxation Implications', L.S.G. 20 Jan. 1988, 19–20.

Lorvellec, L., 'Le régime juridique des transferts de quotas laitiers: commentaire du décret no.87–608 du 31 juillet 1987', [1987] Revue de Droit Rural 409–17.

Moss, J. R., 'Les quotas européens: aspects de l'expérience anglaise et galloise', [1994] Revue de Droit Rural 483–7.

Moss, J. R., Davis, N., and Saunders, J., 'Milk Quota Update: Autumn 1993', (1993) Bulletin of the Agricultural Law Association, Issue 11, 8–10.

Munir, A. E., 'Milk Marketing Upheaval', (1994) 138 Sol.J. 420–1.

Neri, S., 'Le principe de proportionnalité dans la jurisprudence de la Cour relative au droit communautaire agricole', (1981) 4 R.T.D.E. 652–83.

Newth, J., 'Retiring too slowly', (1994) 132 Taxation 304–5.

Pinfold, E., 'What is a Milk Quota?', (1992) 136 Sol.J. 824–5.

Schmitthof, C. M., 'The Doctrines of Proportionality and Non-discrimination', (1977) 2 E.L.Rev. 329–34.

Schockweiler, F., 'Le régime de la responsibilité extra-contractuelle du fait d'actes juridiques dans la Communauté européenne', (1990) 26(1) R.T.D.E. 27–74.

Sharpston, E., 'Legitimate Expectations and Economic Reality', (1990) 15 E.L.Rev. 103–60.

Snape, J., 'Transfers of Milk Quotas: Law and Tax', (1995) 2 Private Client Business 150–61.

Townend, H., 'The Dangers of Permanently Transferring Milk Quota', (1992) Bulletin of the Agricultural Law Association, Issue 7, 2.

Usher, J. A., 'Rights of Property: How Fundamental?', (1980) 5 E.L.Rev. 209–12.

Vadja, C., 'Some Aspects of Judicial Review within the Common Agricultural Policy', (1979) 4 E.L.Rev. 244–61 and 341–55.

Williams, D. and Paskins, H., 'Milk Quota', (1993) 207 Tax Journal 13–15 and 208 *Tax Journal* 13–15.

Wils, W., 'Concurrent Liability of the Community and a Member State', (1992) 17 E.L.Rev. 191–206.

PAPERS

Conway, A. G., 'Milk Quota Review in relation to CAP Objectives, Single EC Market, and GATT Negotiations', *Paper presented to the Centre for European Policy Studies Seminar*, Brussels, 8 March 1989.

Milk Quotas: Law and Practice: Papers from the ICEL Conference — June 1989 (Trinity College, Dublin, 1989).

210 *Milk Quotas*

OFFICIAL PUBLICATIONS

European Community

Court of Auditors: *Special Report No.2/87 on the Quota/Additional Levy System in the Milk Sector*, [1987] OJ C266/1.
Court of Auditors: *Special Report No.4/93 on the Implementation of the Quota System intended to control Milk Production*, [1994] OJ C12/1.
European Commission: *Report on the Agricultural Situation in the Community* (published annually).

United Kingdom

House of Commons: *First Report from the Agriculture Committee: the Implementation of Dairy Quotas*, Session 1984–5, H.C. 14.
House of Commons: *Third Report from the Agriculture Committee: Trading of Quota*, Session 1994–5, H.C.512.
Ministry of Agriculture, Fisheries, and Food: *The Mobility of Quota* (1985).
Ministry of Agriculture, Fisheries, and Food: *Milk Quotas: Proposals to amend the Dairy Produce Quotas Regulations 1993* (1993).

MISCELLANEOUS

Country Landowners Association: *Milk Quotas: M2/89* (London, 1989).
Country Landowners Association: *Milk Quotas: M2/93* (London, 1993).
Federation of UK Milk Marketing Boards: *United Kingdom Dairy Facts and Figures* (published annually, that for 1994 being produced by the England and Wales Residuary Milk Marketing Board) (Thames Ditton, Surrey).
Federation of UK Milk Marketing Boards: *EC Dairy Facts and Figures* (published annually, that for 1994 being produced by the Residuary Milk Marketing Board) (Thames Ditton, Surrey).
Ministry of Agriculture, Fisheries, and Food's CAP Review Group: *European Agriculture: the Case for Radical Reform* (1995).
Ryan-Purcell, O., '*European Community Milk Quota Regulations in Ireland*', (unpublished LLM Thesis, University of Limerick, 1992).

Index